Green Energy and Technology

Climate change, environmental impact and the limited natural resources urge scientific research and novel technical solutions. The monograph series Green Energy and Technology serves as a publishing platform for scientific and technological approaches to "green"—i.e. environmentally friendly and sustainable—technologies. While a focus lies on energy and power supply, it also covers "green" solutions in industrial engineering and engineering design. Green Energy and Technology addresses researchers, advanced students, technical consultants as well as decision makers in industries and politics. Hence, the level of presentation spans from instructional to highly technical.

Indexed in Scopus.

Indexed in Ei Compendex.

Ali Ahmadian • Ali Elkamel • Ali Almansoori
Editors

Carbon Capture, Utilization, and Storage Technologies

Towards More Sustainable Cities

Editors
Ali Ahmadian
Department of Electrical Engineering
University of Bonab
Bonab, Iran

Department of Chemical Engineering
University of Waterloo
Waterloo, ON, Canada

Ali Elkamel
Department of Chemical Engineering
University of Waterloo
Waterloo, ON, Canada

Department of Chemical Engineering
Khalifa University
Abu Dhabi, United Arab Emirates

Ali Almansoori
Department of Chemical Engineering
Khalifa University
Abu Dhabi, United Arab Emirates

ISSN 1865-3529 ISSN 1865-3537 (electronic)
Green Energy and Technology
ISBN 978-3-031-46589-5 ISBN 978-3-031-46590-1 (eBook)
https://doi.org/10.1007/978-3-031-46590-1

This Springer imprint is published by the registered company Springer Nature Switzerland AG
The registered company address is: Gewerbestrasse 11, 6330 Cham, Switzerland

Paper in this product is recyclable.

Preface

Despite the consideration of alternative energy resources and increasing the energy efficiency in the systems to decrease the amount of CO_2 emissions, the cumulative rate of CO_2 in the atmosphere needs to be decreased to limit the detrimental effects of climate change. Therefore, regardless of the extension of clean and more efficient energy systems, carbon-removing technologies need to be implemented. Carbon Capture, Utilization, and Storage (CCUS) is a novel technology that captures CO_2 from facilities including power plants, the transportation systems, and industrial sectors. The CCUS technologies can deliver 'negative emissions' by removing CO_2 directly from the atmosphere or from biomass-based energy and storing the CO_2. Therefore, CCUS technologies need to be implemented in the smart sustainable cities.

This book is an attempt to bring together the experts from the different disciplines related to carbon capture, utilization, and storage process and its impact on sustainable cities development. It contains eight chapters in which numerous researchers and experts from academia and industries are collaborated. The breakdown of the chapters is as follows:

- Chapter 1 describes the important fuels and chemicals and the synthesis methods of each. The use of carbon dioxide in the beverage and food industry is therefore considered. Moreover, the two types of carbon mineralization – in situ and ex situ, which are thought to be the most recent and efficient techniques for carbon utilization – are covered and the applications, products, challenges and risks of each of these techniques are clearly discussed.
- Chapter 2 evaluates the capabilities of CO_2 detection satellites as objective, independent, potential, low-cost and external data sources for monitoring CO_2 emissions from human activities.
- Chapter 3 discusses a much more general framework which allows different capacities for the booster stations. Furthermore, the boosters can be installed at any location, depending on pressure losses along the pipeline.

- Chapter 4 reviews the concept of Power-to-X technologies and the electrification of the chemical industry.
- Chapter 5 provides an overview of machine learning concepts and general model architectures in the context of post-combustion carbon capture. Also, this chapter presents and compares different machine learning models within the field of absorption-based carbon capture. The strengths and limitation of the strategies used in the creation of past models are discussed.
- Chapter 6 presents a design and optimization framework for a tidal power generation plant in the Bay of Fundy, Canada, in order to reduce the operation's cost and emission pollution.
- Chapter 7 presents a systematic framework to integrate renewable energy technologies for the oil and gas industry focusing on solar energy use to meet hydrogen requirements of the crude oil upgrading process for bitumen feedstock in tar sands processing.
- Chapter 8 represents a comprehensive review on CO_2 monitoring satellites.

The editors of the book warmly thank all the contributors for their valuable works. Also, we would like to thank the respected reviewers who improved the quality of the book by the valuable and important comments.

Waterloo, ON, Canada Ali Ahmadian
Waterloo, ON, Canada Ali Elkamel
Abu Dhabi, United Arab Emirates Ali Almansoori

Contents

Chapter 1
Carbon Utilization Technologies & Methods

Reza Mahmoudi Kouhi, Mohammad Milad Jebrailvand Moghaddam, Faramarz Doulati Ardejani, Aida Mirheydari, Soroush Maghsoudy, Fereshte Gholizadeh, and Behrooz Ghobadipour

1.1 Introduction

Carbon utilization is the process of using captured CO_2 as a resource to make value-added products, and it is also an important aspect of climate mitigation. Generally, there are three categories carbon utilization technologies can be divided into: chemical technologies, biological technologies, and mineralization processes (Fig. 1.1).

CO_2 is utilized in chemical processes to produce polymers as well as organic compounds such as acyclic carbonates and cyclic carbonates. The production of energy carriers and transportation fuels such as methanol opens more opportunities for the capturing of CO_2. Liquid fuels are not considered long-term alternatives since they ultimately burn out. In biological technology, microorganisms like algae, cyanobacteria, and proteobacteria are utilized to convert CO_2 into a range of useful chemicals, such as ethylene and ethanol. High-value chemicals may also be produced in the pharmaceutical and food sectors. In the approach like chemical methods, CO_2 is not permanently stored, as it is released back into the atmosphere when the biofuel is burned. But the fuel is a carbon-free product since first it captures carbon from the atmosphere before entering it again by burning. The third group of

R. Mahmoudi Kouhi · M. M. Jebrailvand Moghaddam · F. Doulati Ardejani (✉) ·
A. Mirheydari · S. Maghsoudy · F. Gholizadeh
School of Mining, College of Engineering, University of Tehran, Tehran, Iran

Climate Change Group, Mine Environment & Hydrogeology Research Laboratory (MEHR Lab.), University of Tehran, Tehran, Iran
e-mail: reza_mahmoudi@ut.ac.ir; milad.jebrailvand@ut.ac.ir; fdoulati@ut.ac.ir; aida.mirheidari@ut.ac.ir; s.maghsoudy@ut.ac.ir; fereshtegholizade@ut.ac.ir

B. Ghobadipour
Climate Change Group, Mine Environment & Hydrogeology Research Laboratory (MEHR Lab.), University of Tehran, Tehran, Iran

School of Civil Engineering, Iran University of Science & Technology, Tehran, Iran

A. Ahmadian et al. (eds.), *Carbon Capture, Utilization, and Storage Technologies*, Green Energy and Technology, https://doi.org/10.1007/978-3-031-46590-1_1

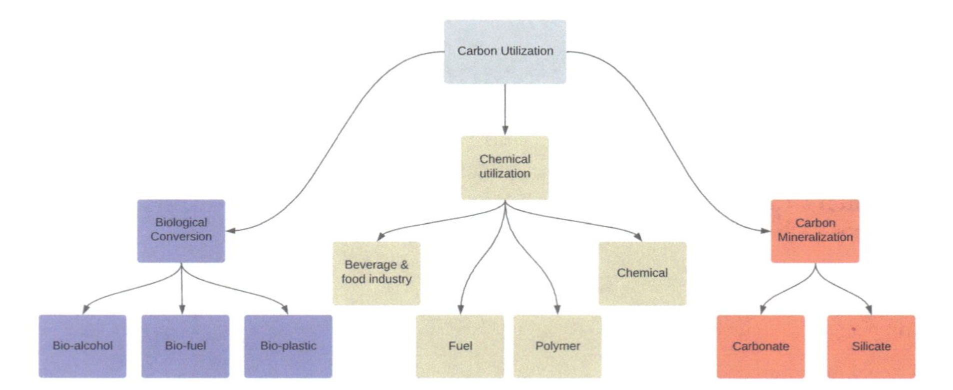

Fig. 1.1 Carbon utilization categories and products

carbon utilization technologies is the mineralization process which is classified into two types: in situ and ex situ mineralization. Mineral trapping or in situ mineralization is underground geological sequestration where a fraction of injected CO_2 reacts with alkaline rocks in the target formation and it forms solid carbonate species. In ex situ mineralization, CO_2 reaction takes place in an industrial process. The final product obtained by this technology can store CO_2 for a long time.

The most important challenges facing carbon utilization technologies are high energy consumption, long-term effects, and the cost of raw materials required. Economic issues regarding different methods, durability over time, and insufficient maturity of the technologies are the other issues that should be considered. One of the most important advantages of using carbon utilization is its ability to be used in sectors that are responsible for around 53% of carbon dioxide emitted into the air (Fig. 1.2). The usage of alternative fuels leads to a reduction in carbon emissions in the transport and electricity and heat sectors. In addition, the construction and industrial sectors reduce their carbon emissions through the manufacturing of carbonates from industrial wastes. Utilization approaches have the potential to reduce about one-fifth of the emissions necessary in the industrial sectors. It is the only option for significantly reducing direct emissions from other industrial point sources, and it will play a significant role in reducing CO_2 emissions from fossil fuel–based power plants. It is estimated that the use of carbon utilization will help cut CO_2 emissions by up to 32% by 2050. Up to 2060, industrial operations may accumulate more than 28 Gt of CO_2, with the chemical, steel, and cement subsectors accounting for the majority of this [2].

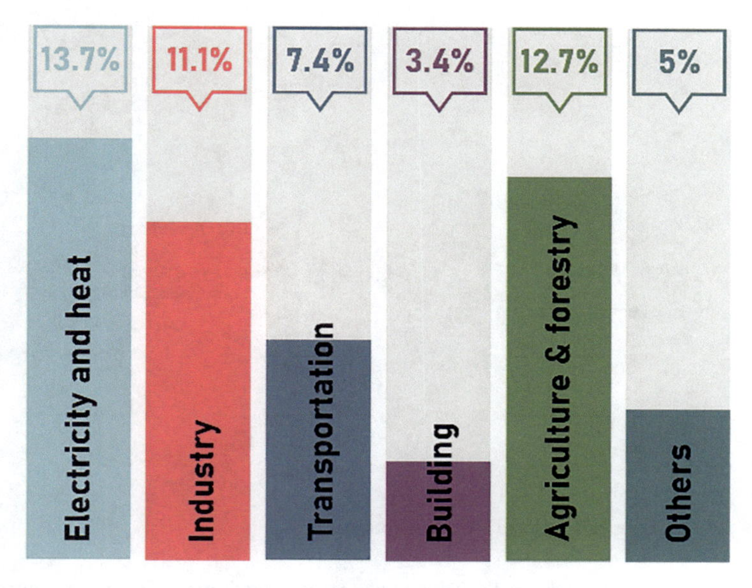

Fig. 1.2 Gross estimate of greenhouse gas emissions by various segments. (Modified after [1])

1.2 CCS Versus CCU

Carbon capture and storage (CCS) and carbon capture and utilization (CCU) refer to technologies that capture CO_2. In CCS methods, CO_2 is permanently stored while the major purpose of CCU is to convert it into valuable products such as fuels and chemicals. Both CCS and CCU are based on carbon capture, but the difference is what happens after the capture phase. Figure 1.3 shows the scope of each of these technologies' effects, as well as their similarities. As can be observed, the method of in situ mineralization is the borders between the usage of CCU and CCS technologies, implying that these two approaches can be classed in both.

1.3 Fuels and Chemicals

The main source of energy used in current energy systems is fossil fuels, which result in the generation of large amounts of carbon dioxide when used in transportation and industry. Therefore, it is necessary to find alternatives for them. Carbon dioxide conversion into fuels and chemicals reduces greenhouse gas emissions and dependence on petrochemicals. The utilization of CO_2 as a feedstock for fuel synthesis as well as chemicals has shown many potential environmental and economic benefits. Several industries, including fuel cells, power plants, and transportation, can utilize the produced fuel. CO_2 is a thermodynamically stable molecule; thus in order to utilize it and produce high fuel yields, a lot of heat and catalyst inventory must be

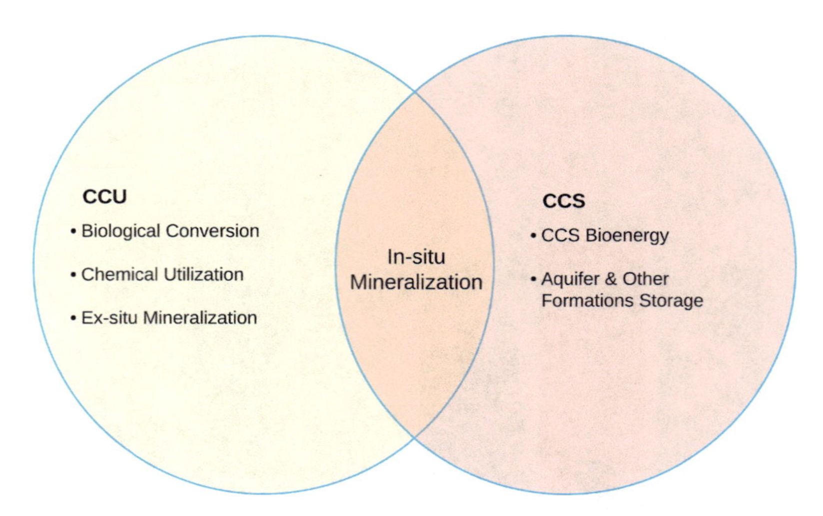

Fig. 1.3 The relations of CCU and CCS technologies

Table 1.1 Main chemicals and fuels that are now manufactured from CO_2 on a worldwide scale [3, 4]

<table>
<tr><th colspan="2">Product</th><th>Production (Mt/year)</th><th>CO_2 utilization (tCO_2/t product)</th><th>Technology readiness level</th></tr>
<tr><td colspan="2">Methane</td><td>1100–1500</td><td>2.750</td><td>CO_2 methanation: 7</td></tr>
<tr><td colspan="2">Methanol</td><td>65.00</td><td>1.373</td><td>Hydrogenation of CO_2: 8–9</td></tr>
<tr><td colspan="2">Formic acid</td><td>1.00</td><td>0.956</td><td>Electrochemical reduction of CO_2: 6</td></tr>
<tr><td colspan="2">Dimethyl ether</td><td>11.40</td><td>1.911</td><td>1–3</td></tr>
<tr><td colspan="2">Liquid fuels</td><td>–</td><td>2.6</td><td>5–9</td></tr>
<tr><td colspan="2">Urea</td><td>180.00</td><td>0.735</td><td>9</td></tr>
<tr><td colspan="2">Salicylic acid</td><td>0.17</td><td>0.319</td><td>9</td></tr>
<tr><td colspan="2">Polycarbonate</td><td>5.00</td><td>0.173</td><td>9</td></tr>
<tr><td colspan="2">Polyurethane</td><td>15.00</td><td>0.300</td><td>8–9</td></tr>
<tr><td rowspan="2">Cyclic carbonates</td><td>Ethylene carbonate</td><td>0.20</td><td>0.499</td><td rowspan="2">4–5</td></tr>
<tr><td>Propylene carbonate</td><td>0.20</td><td>0.431</td></tr>
<tr><td colspan="2">Dimethyl carbonate</td><td>1.60</td><td>1.466</td><td>8–9</td></tr>
</table>

applied. Carbon dioxide can be utilized to produce energy carriers and transportation fuels such as methane, methanol, formic acid, dimethyl ether, carbon monoxide or synthesis gas (syngas), and Fischer-Tropsch fuels. In addition to synthetic fuels, it is also possible to produce various chemicals such as urea, polymers, formic acid, salicylic acid, acyclic carbonates, cyclic carbonates, and fine chemicals such as biotin using carbon dioxide. Table 1.1 summarizes some chemicals and fuels that are currently being manufactured industrially from CO_2.

1.3.1 Methane Production

One of the most significant energy sources is methane (CH_4), which is mostly obtained from natural gas, a fossil fuel source with relatively low costs, and is used to generate heat, power, and value-added chemicals [5]. CO_2 methanation has recently attracted considerable interest, due to its use in Power-to-Gas (PtG) technology and the upgrading of biogas [6]. In order to effectively incorporate renewable energy sources, such as wind and solar energy, into the current energy mix, PtG processes are viewed as a potential and intriguing solution [7]. In this technology, hydrogen generated from surplus renewable energy is chemically changed into methane, which can be stored and transported using the already-existing, highly developed natural gas infrastructure, by reacting with CO_2 [6].

Among the several PtM techniques already in use, catalytic CO_2 hydrogenation (methanation) has received the most attention, and demonstration units are already in operation in a number of nations [5]. At the beginning of the twentieth century, Sabatier and Senderens conducted the first studies of the methanation reaction, also known as the Sabatier reaction. Through this reaction, CO_2 and H_2 are converted into CH_4 and H_2O (Eq. 1.1) [8].

$$CO_2 + 4H_2 \rightarrow CH_4 + 2H_2O, \Delta H = -165 \text{ kJ.mol}^{-1} \tag{1.1}$$

Due to the exothermic nature of this reaction, products with low temperature and high pressure are preferred in terms of thermodynamics [8]. CO_2 hydrogenation can be thought of as a result of combining reverse water gas shift (RWGS) reaction and CO hydrogenation (Eqs. 1.2 and 1.3) [9].

$$CO_2 + H_2 \rightarrow CO + H_2O, \Delta H_r^0 = 41.2 \text{ kJ.mol}^{-1} \tag{1.2}$$

$$CO + 3H_2 \rightarrow CH_4 + H_2O, \Delta H_r^0 = -206.3 \text{ kJ.mol}^{-1} \tag{1.3}$$

Reactors for methanation might be either biological or catalytic (Fig. 1.4). Methanogenic microorganisms function as biocatalysts in biological methanation [9]. A biogas plant's fermenter or a separate bioreactor can be used to conduct this process [10].

Metals from group VIII of the periodic table catalyze the methanation reaction. Ru was shown to be the most active metal catalyst, followed by Fe, Ni, and Co. Ni is typically chosen as the active component because of its high selectivity and reactivity, and because it is reasonably priced [11]. Despite having advantages over Ni systems, Ru catalysts are more expensive. Given the low cost and wide availability of methane from natural gas, hydrogenation of CO_2 to methane is not now feasible on a big scale and is not anticipated to be in the near future. Furthermore, methane has a significantly lower economic value than the conversion of CO_2 into a variety of other compounds [12].

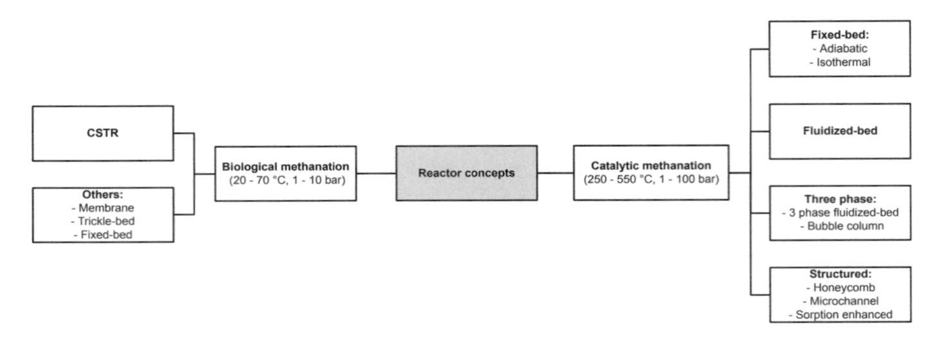

Fig. 1.4 Concepts for reactors that produce substitute natural gas [9]

The electrochemical reduction of CO_2 is another potential method for producing methane from CO_2. This technique is still being validated in the lab. However, recent results have emphasized the attractive characteristics of this path [5]. Currently, only copper is capable of catalyzing the conversion of CO_2 into hydrocarbons, particularly methane, in an aqueous solution. Higher overpotentials, low activity, and poor product selectivity are problems with conventional Cu electrodes [13]. To create catalysts with improved methane selectivity, more research is still required.

1.3.2 Methanol Production

The most basic liquid hydrocarbon that can be used as a fuel, a hydrogen carrier, or a feedstock for creating more intricate chemical compounds is methanol (CH_3OH) [14]. Formaldehyde, acetic acid, dimethyl ether (DME), and methyl tertiary-butyl ether (MTBE) are the primary chemical derivatives of methanol [15]. The methanol-to-olefins process creates light olefins like ethylene and propylene, which can be utilized to make polymers and hydrocarbon fuels. Additionally, methanol is converted into dimethyl carbonate in supercritical CO_2, which is a helpful intermediary for derivatives utilized in polycarbonates and polyurethanes [16].

According to Eq. (1.4), syngas, which has a CO/H_2 mixture, is being used to create methanol on an industrial scale. Currently, syngas (mixture of CO and H_2) produced mostly from natural gas reforming is transformed into methanol at temperatures between 250 and 300 °C and pressures between 5 and 10 MPa, using a $CuO/ZnO/Al_2O_3$ catalyst [15, 17].

$$CO + 2H_2 \rightarrow CH_3OH, \Delta H = -90.6\ kJ.mol^{-1} \quad (1.4)$$

Currently, a little amount of CO_2 (up to 30%) is typically added to the syngas. The energy balance and methanol yield both considerably increase with the addition of CO_2 to the CO/H_2 feed. Syngas is low in hydrogen and high in carbon oxides (CO and CO_2). The CO in syngas is transformed to CO_2 via the water-gas shift

(WGS) reaction to increase its H_2 content and promote methanol synthesis (Eq. 1.5) [16].

$$CO + H_2O \rightarrow CO_2 + H_2, \Delta H^0_{298} = -41.2 \text{ kJ mol}^{-1} \tag{1.5}$$

The catalytic hydrogenation process shown in Eq. (1.6) is the most direct method for producing methanol from CO_2 and involves the production of H_2 using water electrolysis, ideally with the use of renewable energy, and the subsequent combination with CO_2 waste streams to create methanol, which is known as the Power-to-Methanol process. This process involves the RWGS (Eq. 1.7) as a secondary reaction and is less exothermic than the syngas-based approach. RWGS reaction is regarded unfavorable since it consumes H_2 and reduces the yield of methanol synthesis. It was discovered that the rate of the direct methanol synthesis from CO_2 was inhibited by the water produced as a byproduct. [5, 15, 16].

$$CO_2 + 3H_2 \rightarrow CH_3OH + H_2O, \Delta H = -49.5 \text{ kJ.mol}^{-1} \tag{1.6}$$

$$CO_2 + H_2 \rightarrow CO + H_2O, \Delta H^0_r = 41.2 \text{ kJ.mol}^{-1} \tag{1.7}$$

Hydrogenation of carbon dioxide to methanol is an efficient CO_2 utilization technique and is considered an effective sustainable development strategy. This method is technically comparable to the production of methanol from syngas for industrial use [16]. If direct hydrogenation of CO_2 to methanol is replaced with methanol production from syngas, improved catalysts are greatly needed [12]. In comparison to conventional synthesis, this method has a better water footprint, but still lacks competitive economic viability [4].

The electrochemical reduction of CO_2 using protons and electrons as a source of H_2 is another method for producing methanol. Due to its complicated kinetics, this reaction requires efficient electrocatalysts. One of the most effective materials for the electrochemical conversion of CO_2 into alcohols, including methanol, has been recognized to be copper or copper-based electrodes. In order to improve the electrochemical CO_2 reduction to CH_3OH, the usage of copper alloys has also been studied. Cu-Zn mixed oxides make up the majority of commercial catalysts used today to produce methanol, demonstrating the metals' synergistic influence on methanol synthesis [5, 12].

1.3.3 Dimethyl Ether (DME) Production

The simplest ether is dimethyl ether (DME), which has the chemical formula CH_3OCH_3. DME has physical properties similar to liquefied petroleum gases (LPG) such as propane and butane. DME has been marketed as a diesel substitute since the mid-1990s. With a high cetane number (55-60), DME has several desirable characteristics over conventional fuels, including very low emissions of pollutants (SO_x, NO_x, CO, and particulate matter) [18, 19].

Indirect synthesis (two-stage) and direct synthesis from syngas (single-stage) are typically the two methods used to produce DME. In the single-stage method, DME is prepared directly from syngas in a single reactor [20]. Fixed-bed reactors have been used for the majority of theoretical studies on single-step DME production [21]. In the two-step process, syngas is first transformed into methanol (Eq. 1.8), which is then dehydrated to produce dimethyl ether (Eq. 1.9). Zeolites and Al_2O_3, in particular, have been suggested as acid catalysts for the dehydration of methanol to DME [22]. In a reactor, WGS reaction can occur concurrently (Eq. 1.10) [19].

$$\text{Methanol synthesis}: CO + 2H_2 \rightarrow CH_3OH, \Delta H = -90.6\ \text{kJ.mol}^{-1} \tag{1.8}$$

$$\text{Methanol dehydration}: 2CH_3OH \rightarrow CH_3OHCH_3 + H_2O, \Delta H = -23.41\ \text{kJ.mol}^{-1} \tag{1.9}$$

$$\text{WGS}: CO + H_2O \rightarrow CO_2 + H_2, \Delta H^0_{298} = -41.2\ \text{kJ mol}^{-1} \tag{1.10}$$

While the current technologies for both methods rely on fossil-based syngas, which again causes environmental issues, recent studies examine the possibility of replacing syngas with CO_2/H_2 feed (Eqs. 1.11 to 1.13) [22].

$$CO_2\ \text{hydrogenetion}: CO_2 + 3H_2 \rightarrow CH_3OH + H_2O, \Delta H = -49.5\ \text{kJ.mol}^{-1} \tag{1.11}$$

$$\text{RWGS}: CO_2 + H_2 \rightarrow CO + H_2O, \Delta H^0_r = 41.2\ \text{kJ.mol}^{-1} \tag{1.12}$$

$$\text{Methanol dehydration}: 2CH_3OH \rightarrow CH_3OHCH_3 + H_2O, \Delta H = -23.41\ \text{kJ.mol}^{-1} \tag{1.13}$$

The direct synthesis of DME from concentrated CO_2 and H_2 has lately gained attention due to the growing interest in CO_2 capture and valorization. The synthesis of methanol is a recognized thermodynamically limited process. As a result, using methanol immediately to create DME via a direct method has the advantageous effect of pushing the equilibrium toward higher conversions. Because of the water forming in greater quantities and the consequently more stringent thermodynamic constraints, the CO_2 to DME process is more difficult than the syngas method and hence necessitates focused attention. A strategy that has been introduced to solve this problem is the in situ removal of water produced in all individual reactions using a membrane reactor [22].

1.3.4 Formic Acid Production

Formic acid (HCOOH) serves as a platform for chemical energy storage in addition to being a valuable chemical that is frequently used as a preservative and antibacterial agent. Through its decomposition to CO_2 and H_2 and potential for reversible transition back to formic acid, this acid is a known hydrogen storage component [18]. Formic acid and its salts have a wide range of uses, including as a starting chemical for esters, alcohols, or medicinal products, as well as in the production of textiles, leather, and dyes and as a cleaning or disinfection solution [23].

Formic acid is produced industrially most frequently via a two-step process: In the first step, methyl formate is generated from methanol and CO (Eq. 1.14), and in the second step, methyl formate is hydrolyzed into formic acid (Eq. 1.15). The second step is thermodynamically unfavorable [5].

$$CH_3OH + CO \rightarrow CH_3COOH, \Delta H_r = -29\ kJ.mol^{-1} \quad (1.14)$$

$$CH_3COOH + H_2O \rightarrow HCO_2H + CH_3OH, \Delta H_r = 16.3\ kJ.mol^{-1} \quad (1.15)$$

Also, formic acid can be produced through the hydrogenation of carbon dioxide (Eq. 1.16). As a result of the conversion of gases into liquids during this process, the reaction is entropically unfavorable. The reaction is therefore exergonic in the aqueous phase and endergonic in the gas phase. However, when the reaction is carried out in the aqueous phase, the presence of the solvent can change the reaction thermodynamics and makes it slightly exergonic (Eq. 1.17). By employing additives, such as specific bases like ammonia (Eq. 1.18) and triethylamine, the equilibrium can be changed in favor of the product. Carbonates, bicarbonates, and hydroxides are frequently used for the reaction in water [24, 25].

$$CO_2(g) + H_2(g) \rightarrow HCO_2H(l), \Delta G^0_{298K} = 32.9\ kJ.mol^{-1} \quad (1.16)$$

$$CO_2(aq) + H_2(aq) \rightarrow HCO_2H(aq), \Delta G^0_{298K} = -4\ kJ.mol^{-1} \quad (1.17)$$

$$CO_2(g) + H_2(g) + NH_3(aq) \rightarrow HCO_2^-(aq) + NH_4^+(aq), \Delta G^0_{298K} = -9.5\ kJ.mol^{-1} \quad (1.18)$$

Numerous homogeneous and heterogeneous catalysts have been developed for CO_2 hydrogenation to formic acid on a lab scale. Transition metal complexes, especially those based on Ir and Ru, have been used in a tremendous amount of attempts, and the results are very remarkable. To become potentially practical, these catalysts require further improvements in selectivity to formic acid and stability. Heterogeneous catalysts, on the other hand, are less studied for this reaction; however, recently the number of examples has notably increased. The heterogeneous catalysts are characterized as follows, with clear practical advantages for continuous operation and product separation: heterogenized molecular catalysts and unsupported and supported bulk/nanometal catalysts [18].

Because of the high market value and widespread use of formic acid, direct electrochemical reduction of carbon dioxide to this substance has emerged as a viable option. This procedure involves supplying electricity to an electrolytic cell. An electrolyte cell is made up of an anode and a cathode with catalyst-coated surfaces, as well as an electrolyte(s) that allows ions to be transferred between the electrodes. Eqs. (1.19) and (1.20) show half-reactions that take place at the anode and cathode of an electrolytic cell set up to make formic acid from CO_2.

$$\text{Cathode} : \text{CO} + 4\text{H}^+ + 4\text{e}^- \rightarrow 2\text{HCOOH} \tag{1.19}$$

$$\text{Anode} : 2H_2\text{O} \rightarrow \text{O}_2 + 4\text{H}^+ + 4\text{e}^- \tag{1.20}$$

The typical operating conditions of this process are ambient temperature and pressure, which is one of its main advantages. However, the primary hurdles for the development of this method are significant overpotentials and limited product selectivity. Various catalysts based on Co, Pb, Pd, Sn, and In metal-free nitrogen-doped carbon materials have been reported for this process over the last few decades [4, 5, 26].

1.3.5 Carbon Monoxide – Syngas Production

Carbon monoxide (CO) is an important chemical product precursor (Fig. 1.5) [27]. Synthesis gas, also known as syngas, is a gaseous fuel mixture of carbon monoxide and hydrogen that is fed to a number of industrial processes, including the direct DME (dimethyl ether) synthesis, the Fischer-Tropsch (F-T) synthesis, the ammonia synthesis, the methanol synthesis, the power and heat generation

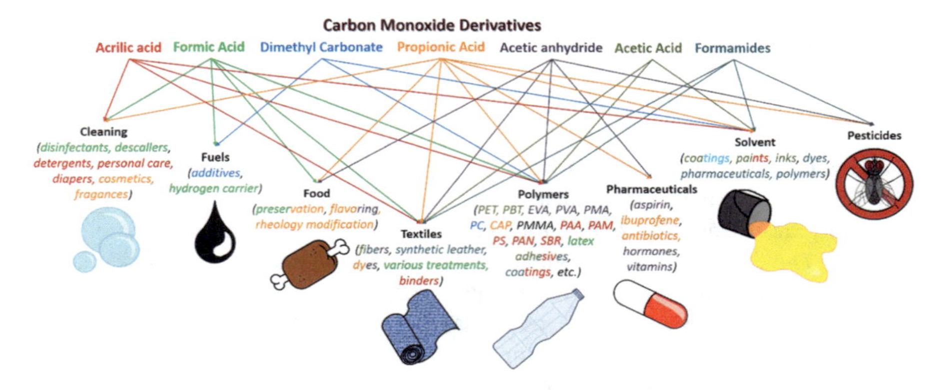

Fig. 1.5 Applications and principal derivatives of carbon monoxide [27]

processes, and the SNG (substitute natural gas) synthesis [28]. Due to its superior efficiency than the direct conversion technologies now in use, syngas remains the industrially favored technology for the indirect conversion of natural gas into higher-value chemicals and fuels for the time being. Although almost any raw material containing carbon can be utilized to produce H_2/CO mixtures, natural gas, liquid hydrocarbon sources, solid fossil carbon sources like coal or lignite, or raw materials obtained from renewable sources are now the most preferred sources [29]. Methane/natural gas is the most extensively utilized raw material for synthesis gas due to its availability, gas composition, and inexpensive cost [30].

Steam methane reforming (SMR), dry methane reforming (DRM), autothermal reforming (ATR), partial oxidation (POX), bireforming (BR), tri-reforming (TR), and combined reforming (CR) have traditionally been used to produce syngas from fossil-based natural gas and coal [28]. When methane is used to create syngas, the process involves the employment of an oxidizing agent that oxidizes methane to carbon monoxide while also creating hydrogen in a ratio that varies depending on the oxidant type. Carbon dioxide is able to function as an oxidizing agent through a procedure called dry reforming [31]. Because DRM is a highly endothermic reaction (Eq. 1.21), equilibrium conversion to syngas must occur at extremely high temperatures [32].

$$CH_4 + CO_2 \rightarrow 2CO + 2H_2, \Delta H^0_{298} = 248 \text{ kJ mol}^{-1} \tag{1.21}$$

The methane dry reforming process is the most endothermic reaction when compared to SMR and ATR [33]. DRM yields syngas with a H_2 to CO ratio that is more compatible with some downstream synthesis processes, such as Fischer-Tropsch synthesis [17].

Due to the difficulty in developing catalysts with a long life-span on stream at a low price acceptable for profit-oriented commercialization, despite its economic and environmental potential, DRM is still in its infancy [34]. The formation of coke and sintering, which quickly deactivate the catalysts, is the main obstacle inhibiting the widespread use of DRM in the industry [32]. It is expected that coke will deposit on the reforming catalyst due to high working temperatures, which increase the molecular energy enough to split the C-H bonds in methane [33]. In order to be used on a large scale in industrial applications, the ideal DRM catalyst must be extremely stable and have better resistance to coke formation. Numerous experiments using supported metal catalysts and noble (ruthenium, rhodium, platinum, palladium, and iridium) and non-noble metals (nickel and cobalt) have been conducted [32].

The dry reforming reaction equilibrium is usually influenced by the co-occurrence of the RWGS reaction (Eq. 1.22) [30].

$$CO_2 + H_2 \rightarrow CO + H_2O, \Delta H^0_{298} = 41.2 \text{ kJ mol}^{-1} \tag{1.22}$$

The H_2/CO molar ratio is decreased as a result of the RWGS reaction by consuming H_2 [35]. It is an endothermic reaction, so formation of CO is favored at high temperatures [36]. Only in the presence of a suitable and sustainable source of hydrogen and thermal energy at the proper temperature level the RWGS reaction will be commercially attractive as a source for syngas [29]. For this reaction, a variety of heterogeneous catalysts have been utilized, including systems based on copper, iron, or ceria (Cerium (IV) oxide). However, in general, they have low thermal stability, and methane commonly forms as an unfavorable byproduct [12]. In designing a suitable catalyst for the RWGS reaction, criteria of high activity and high CO selectivity should be considered [36].

The direct electrolysis of carbon dioxide to carbon monoxide and oxygen is another method for producing CO from carbon dioxide [37]. Three electrolysis techniques are used in this procedure: solid oxide electrolysis at high temperature, molten carbonate electrolysis, and low temperature electrolysis using a solution-phase or gas diffusion electrolysis cell. The only CO_2 electrolysis method that is nearing commercialization is high-temperature electrolysis in solid oxide cells [38].

1.3.6 Liquid Hydrocarbons Production (Fischer-Tropsch)

A good substitute for storing renewable energy is liquid hydrocarbons. They are the main source of energy for use in aviation and transportation [20]. Carbon dioxide can also be converted to hydrocarbons through Fischer-Tropsch (FT) and methanol pathways. For the FT pathway, the intermediate product is CO (or a synthesis gas), while for the methanol pathway, it is methanol [39]. There are three steps in both pathways [17]:

- Using renewable electricity to electrolyze water to produce hydrogen.
- Conversion of CO_2 to an intermediate product, methanol or CO.
- Liquid hydrocarbon synthesis, followed by improvement or conversion to the desired fuel.

Synthesis gas can be converted into a variety of products, including synthetic fuels, lubricants, and petrochemicals, using the FT process [40]. In the Fischer-Tropsch pathway, RWGS reaction (Eq. 1.23) is used to produce syngas, which is then converted to liquid hydrocarbons via the Fischer-Tropsch reaction [39]. Synthesis of alkanes, as the main products of FT processes, alkenes, and alcohols are given in Eqs. (1.24) through (1.26) [4]. Ni, Fe, and Cu catalysts can be used in the RWGS reaction; also, Co, Fe, and Ru catalysts can be used in the Fischer-Tropsch synthesis, respectively [39].

$$CO_2 + H_2 \rightarrow CO + H_2O \tag{1.23}$$

$$(2n + 1)H_2 + nCO \rightarrow C_nH_{2n+2} + nH_2O \tag{1.24}$$

$$2nH_2 + nCO \rightarrow C_nH_{2n} + nH_2O \tag{1.25}$$

$$2nH_2 + nCO \rightarrow C_nH_{2n+2}O + (n-1)H_2O \quad (1.26)$$

In the methanol pathway, CO_2 and H_2 react over a metallic catalyst to produce methanol, which is then converted into other hydrocarbons over acidic catalysts [39]. Through a series of reactions, including DME synthesis, olefin synthesis, oligomerization, and hydrotreating, methanol is transformed into gasoline, diesel, and kerosene [17].

Currently, methanol is generated from synthesis gas using a Cu-ZnO-Al_2O_3 catalyst (Eq. 1.27). Recent research efforts have concentrated on the development of catalysts that support the direct conversion of CO_2 to methanol (Eq. 1.28). It is vital to utilize a very selective catalyst for this reaction because it is favored at low temperatures and high pressure and can yield a variety of byproducts [39].

$$CO + 2H_2 \rightarrow CH_3OH, \Delta H_r^{298k} = -90.6\ kJ.mol^{-1} \quad (1.27)$$

$$CO_2 + 3H_2 \rightarrow CH_3OH + H_2O, \Delta H_r^{298k} = -49.5\ kJ.mol^{-1} \quad (1.28)$$

Another way to create fuel-like hydrocarbons that can be used in the current infrastructure is through electroreduction of CO_2 [41]. There are a number of systems that can produce products with new carbon-carbon bonds, even though the reduction of CO_2 to C1 feedstocks such CO, methane, formic acid, or methanol is the process that occurs most frequently [12]. Although the Faradaic efficiency is still low due to H_2O dissociation to H_2, Cu-based electrodes are perfectly suitable in activating CO_2 [41]. As mentioned above, the electroreduction of CO_2 to value-added compounds shows promise, but is still far from commercialization due to the high overpotential of this reaction and the low activity of the currently available catalysts [42].

1.3.7 Urea Production

Another non-toxic product made from carbon dioxide is urea (CH_4N_2O). Liquid and solid fertilizers, urea-formaldehyde resins used to manufacture adhesives and binders, melamine for resins, livestock feeds, NO_x control from boilers and furnaces, and a variety of chemical applications are all the uses of urea [43].

Reforming natural gas to produce ammonia and carbon dioxide is the most widely used process for producing urea [44]. The production of urea results from the reaction of carbon dioxide and ammonia at a temperature between 185 and 190 °C and a pressure between 180 and 200 atm. Two equilibrium reactions known as Basaroff reactions with incomplete reactants conversion are involved in this process: Ammonium carbamate (H_2N-$COONH_4$) is generated in the first stage by the fast and exothermic reaction of liquid ammonia with gaseous CO_2 at high temperature and pressure (Eq. 1.29). In the next step, ammonium carbamate decomposes slowly

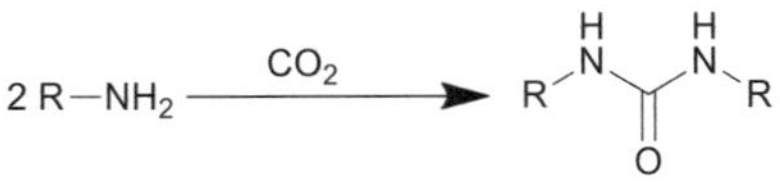

$$2\,R{-}NH_2 \xrightarrow{CO_2} R{-}NH{-}C(=O){-}NH{-}R$$

Fig. 1.6 Urea derivatives synthesis from amine and CO_2 [47]

and endothermically into urea and water using the heat produced by previous reaction (Eq. 1.30) [45, 46].

$$2NH_3 + CO_2 \rightarrow NH_2COONH_4, \Delta H = -117\ \text{kJ.mol}^{-1} \quad (1.29)$$

$$NH_2COONH_4 \rightarrow NH_2CONH_2 + H_2O, \Delta H = 15.5\ \text{kJ.mol}^{-1} \quad (1.30)$$

The use of CO_2 in the synthesis of urea derivatives has received a lot of interest. Anti-cancer agents, plastic additives, gasoline antioxidants, agricultural pesticides, dyes, medicines, gasoline antioxidants, and corrosion inhibitors are just a few uses for urea derivatives. The traditional process for producing urea derivatives includes the reaction of amines with phosgene, carbon monoxide, or isocyanate, which has serious toxicological and environmental issues. One of the main aims of Green Chemistry nowadays is to replace these dangerous reagents in chemical processes. As a result, there has been a significant advancement in the production of urea derivatives through the reaction of amines with CO_2 either with or without the use of a dehydrating agent, using basic ionic liquids or base catalysts [47–49] (Fig. 1.6).

1.3.8 *Polymers*

A unique class of chemicals known as polymers is employed in the manufacturing process for plastics and resins. Polymers, such as polyurethanes and polycarbonates, are adaptable materials with several practical uses, including those in the electrical and electronic industries, the automobile sector, packaging, the medical industry, personal care goods, and the construction [50]. Up until this point, the primary raw materials used in the manufacturing of polymers were petrochemicals[51]. However, the chemical industry is under pressure to discover practical substitutes for the manufacture of renewable chemicals and polymers due to the depletion of fossil fuels and the legal demand for sustainable and renewable plastics under the circular economy [50]. As a raw material for the synthesis of polymers, CO_2 can partially replace petrochemicals. One example is the copolymerization of epoxides with CO_2 to create polycarbonates [17]. As potential, more environmentally acceptable raw materials for plastics, CO_2-based polymers have received a lot of industrial interest [52]. Additionally, using CO_2 to produce different biodegradable polymers is seen to be a cost-effective strategy from an economic perspective [20]. There are two chemical methods for including CO_2 in the production of polymers: direct and indirect methods. Both strategies have been shown to be feasible and possible [48, 49].

1.3.8.1 The Direct Method

The direct method produces high CO_2 content polymers such as polycarbonates, polyols, polyurethanes, polyureas, and polyesters by using CO_2 as a monomer in combination with proper reagents and catalysts [12].

1.3.8.1.1 Polycarbonates (PCs) from CO_2

Aromatic PCs are utilized as engineering plastics in automobiles, electrical and electronic equipment, and construction because of their great impact resistance, stiffness, toughness, superior thermal stability, transparency, and flame retardancy. The toxic and destructive phosgene reaction with 1,2-diol is the traditional method for producing polycarbonates. The copolymerization of epoxides, such as propylene oxide, cyclohexene oxide, vinyl oxide, ethylene oxide, and styrene oxide and CO_2, is an alternate method for the selective production of PCs. This process is the most promising application of CO_2. In general, transition metals or metals from the main group of elements, such as cobalt, zinc, chromium, magnesium, and aluminum, are used as homogeneous or heterogeneous catalysts for the copolymerization of CO_2 and epoxides. Compared to heterogeneous catalysts, homogeneous catalysts are more active and selective. Current CO_2 copolymerization research focuses on the development of catalysts for the production of polymers with tailored properties and derived from renewable epoxides such as limonene oxide, cyclohexadiene oxide, and α-pinene oxide [17, 51].

1.3.8.1.2 Polyurethanes (PUs) from CO_2

Polyurethanes (PUs), one of the most significant polymers, are used in a variety of products in daily life, including adhesives, sealants, coatings, elastomers and foams, heart valves, and cardiovascular catheters. They are manufactured commercially using polyaddition of diisocyanates with di- or polyols. Establishing isocyanate-free production methods has received recent attention in the field of PUs; CO_2 can play a significant role in this vital transition. When CO_2 reacts with cyclic amines like aziridines and azetidines or the N-analogs of epoxides, PUs can be produced [50].

1.3.8.1.3 Polyureas (PUA) from CO_2

Polyureas (PUAs) are polymers with urea linkages built into their backbone. They are used as linings, joint sealants, and microcapsules among other things in a variety of industries, including the building industry, the automobile industry, household products, and marine-related technology. PUAs are created commercially by the polyaddition process utilizing the reagents diisocyanate and diamine. These polymers can be made via non-isocyanate methods using CO_2-sourced (a)cyclic carbonates or urea, or direct CO_2 copolymerizing with diamines [50].

1.3.8.2 The Indirect Method

The indirect method involves converting CO_2 into a different monomer, such as methanol, ethylene, carbon monoxide, organic carbonates, dimethyl carbonate, or urea, which enables the synthesis of a wide range of polymers with a variety of controlled and specified properties. Additionally, CO_2 can be used to create chemical building blocks for polymer synthesis, specifically urea. This makes it possible to create a variety of thermosetting polymers, including Melamine-Formaldehyde (MF) and Urea-Formaldehyde (UF) resins, as well as commercial plastics like Polyoxymethylene (POM) or Polymethylmethacrylate (PMMA) [51].

1.3.9 Other Chemicals

In addition to urea and polymers, the production of other chemicals, such as salicylic acid, inorganic and organic carbonates, fine chemicals such as biotin, etc., is possible by utilizing carbon dioxide. Acyclic (linear) carbonates (e.g., dimethyl carbonate [DMC], diethyl carbonate [DEC], diallyl carbonate [DAC], and diphenyl carbonate [DPC]) and cyclic carbonates (e.g., ethylene carbonate [EC], cyclohexene carbonate [CC], propylene carbonate [PC], and styrene carbonate [SC]) make up the majority of the organic carbonates class [53]. CO_2 and two equivalents of an alcohol, such as methanol, can be used to produce linear carbonates directly. Linear carbonates are used as solvents, reagents (for alkylation or acylation reactions), and gasoline additives. The cyclic carbonates can be produced by reacting CO_2 with a cyclic ether (e.g., an epoxide) or a diol. They are used as monomers for polymers, components of special materials, and also in the synthesis of hydroxyesters and hydroxyamines [45, 53].

1.3.10 Beverage and Food Industry

Food production is possible using CO_2 that is captured for CCU. The principal applications for food-grade CO_2 at the moment are the creation of carbonated beverages, deoxygenated water, milk products, and food preservation. In addition to serving as a carbonating agent for the creation of champagne, alcoholic drinks, and soft drinks, carbon dioxide can also be utilized as a preservative, packing gas, and flavor solvent. Potential CO_2 merchant markets in the US require between 3.2 and 4.0 million metric tons of CO_2 annually for food processing and between 1.6 and 2.4 million metric tons of CO_2 annually for carbonated beverages. CO_2 is utilized to

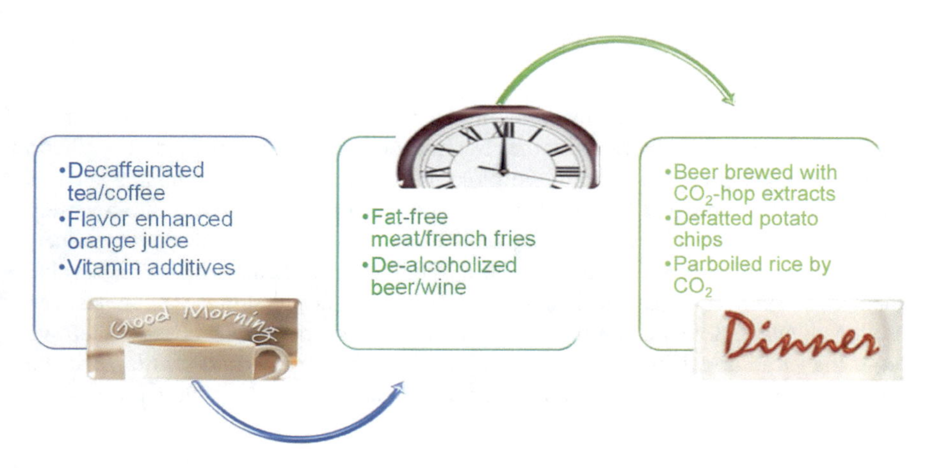

Fig. 1.7 Typical food items obtained through SFE [52]

prevent food from oxidizing. Although N_2 gas is frequently used to prevent oxidation, CO_2 and N_2 together are preferable for antioxidative food packaging. Additionally, antibacterial behavior of CO_2 has been demonstrated in a variety of literature. Food freshness is preserved as a result, extending its shelf life [20, 52, 54].

Mechanical refrigeration is mostly employed during transportation and storage in traditional food preservation. However, liquid carbon dioxide, dry ice (i.e., the solid form of CO_2), and modified atmosphere packaging (MAP) technologies are more frequently employed for refrigeration of foods that need freeze drying (dehydration). CO_2 is frequently used as a flushing gas in MAP. Because of its high solubility in food matrices, the presence of carbon dioxide in the atmosphere package may reduce the pressure or volume of package, so balancing the pressure between the inside and outside of the package. To prevent high CO_2 dissolution into foods, the CO_2-based MAP strategy should be implemented with extreme professionalism in accordance with food attributes and operational requirements. High levels of dissolved CO_2 cause packaging to collapse and produce products with a poor texture and flavor [20, 52].

Supercritical fluid extraction (SFE) technology is a method for utilizing CO_2 in flavors as well as coffee decaffeination, which is advantageous for the separation and extraction of heat-sensitive, volatile, and oxidizable components. Compared to traditional separation methods, this method has several advantages, including non-toxicity, non-corrosiveness, and chemical stability of the extraction agent in SFE, as well as its reusability after decompression, controllability of SFE extraction capability by adjusting the main operating factors, and providing better permeability compared to other solvent approaches. Due to the aforementioned benefits, supercritical CO_2 extraction (SCE) technology is preferred in the food processing industry. As seen in Fig. 1.7, this technology is currently used widely in daily life [52].

1.4 Biological Conversion

The utilization of microorganisms to produce a variety of products is known as biological conversion of CO_2. In some circumstances, the emerging field of synthetic biology has the potential to improve biological systems. Microorganisms such as algae, cyanobacteria, and β-Proteobacteria take up CO_2 and convert it into a variety of valuable compounds during biological CO_2 conversion. Some of these products could be large-scale bulk chemicals like ethylene and ethanol. More high-value chemicals, such as medicines, nutrition, cosmetics, and fragrances, can also be produced; while low in volume, these items may give a more cost-competitive route than traditional industrial synthesis routes [55]. In this part, we look at the microorganisms utilized in biological conversion and the products they produce.

1.4.1 Microorganisms

In this section, we look at the key microorganisms used in biological conversion like algae, cyanobacteria, and β-Proteobacteria that have received the most interest and could potentially be turned into industrial-scale bioprocesses.

1.4.1.1 Algae

Algae are a wide category of aquatic eukaryotic organisms that can do photosynthesis. Its primary habitats include moist, wooded places, still waters, lakes, and pools. Algae are commonly classified into two types based on their size and shape: macroalgae and microalgae. Similar to kelps, algae are composed of many cells that join together to form structures such as roots and stems, as well as the leaves of more mature plants. The great majority of microalgae or microscopic photosynthetic creatures are present in unicellular form and can be found in a wide range of environments. Microalgae are regarded to be one of the earth's oldest life forms. They can thrive in a number of natural habitats, including freshwater, brackish water, and seawater and can adapt to a variety of high temperatures and pH levels. On the basis of their habitats and physical characteristics, microalgal species can also be categorized further. These groups include euglenoids, diatoms, green algae (Chlorophyceae), red algae (Rhodophyceae), yellow-green algae (Xanthophyceae), golden algae (Chrysophyceae), and Chlorophyceae (green algae) [56].

The Calvin-Benson-Bassham (CBB) cycle allows algae to utilize CO_2. The CBB cycle, in fact, is an essential biological mechanism for converting CO_2 from the atmosphere to organic matter. The main enzyme for CO_2 fixation in this cycle is ribulose-1,5-bisphosphate carboxylase/oxygenase (RuBisCO). Aside from the CBB cycle, nature has identified five other carbon fixation mechanisms, the most efficient of which is the reductive acetyl-CoA process under anaerobic conditions [57]. For their ability to fix inorganic carbon, both macro- and microalgae are investigated and utilized. Their potential is attributed to their widespread distribution (especially in moist conditions), high biomass capability, rapid CO_2 uptake and utilization, and, most crucially, their ability to make secondary products with high commercial value from biomass. The most industrially important component of algal biomass is lipid, which is used to make secondary goods such as biofuels and lubricants. To maximize the value of algal carbon capture and utilization, it is critical to select high lipid-producing strains and optimize growing parameters such as light, temperature, and pH [58].

1.4.1.2 Cyanobacteria

Cyanobacteria (or blue-green algae) are phylogenetically a group of Gram-negative photosynthetic prokaryotes having widespread distribution ranging from hot springs to the Antarctic and Arctic regions. The role of cyanobacteria in nitrogen fixation and in the maintenance of the fertility of rice is well documented. [59] Additionally, they are believed to have contributed to the early rise in atmospheric O_2 and the lowering of CO_2 around 2.3 billion years ago. 20–30% of Earth's primary photosynthetic productivity is accounted for by cyanobacteria, which convert solar energy into chemical energy stored in biomass at a rate of 450 TW [60, 61]. RuBisCO, which catalyzes the same reaction as in the CBB cycle in algae, is in charge of the carbon utilization in cyanobacteria. Due to their simpler structure than algae, cyanobacteria are more effective in fixing carbon from the atmosphere. However, they cannot produce the same amount of biomass [58, 62].

1.4.1.3 β-Proteobacteria

β-Proteobacteria are a class of Gram-negative bacteria, and one of the eight classes of the phylum Pseudomonadota. Ralstonia Eutropha H16 is a Gram-negative lithoautotrophic bacterium from the Proteobacteria-subclass. It is a common inhabitant of freshwater and soil biotopes and is highly adapted to survive in environments with intermittent anoxia [63]. R.Eutropha lives on hydrogen (H_2) as its only energy source when there are no organic materials present, fixing CO_2 through the CBB cycle. In addition, it is capable of utilizing a wide array of carbon sources for growth and polymer biosynthesis, including sugars, organic acids, fatty acids, and CO_2. The biggest advantage of working with R.Eutropha is the ability to store carbon within its cytoplasm in the form of polyhydroxyalkanoates (PHAs), also known as bio-plastics. Genetic engineering, on the other hand, can be utilized to create polymers of varying lengths. R.Eeutropha is also sought after for its various carbon utilization routes and biocompatibility in the production of pharmaceutical chemicals [58, 64].

1.4.1.4 Other Microorganisms

Other microorganisms, in addition to those mentioned, have the ability to absorb carbon and produce fuel and other valuable industrial substances. For example, acetogenic bacteria such as Clostridium autoethanogenum have the ability to grow and convert CO_2 and CO into low-carbon fuels and chemicals like ethanol, acetone, and butanol [65]. Besides that, there are many microorganisms from an archaeal domain that can fix carbon dioxide through CO_2 fixing pathways [66].

1.4.2 Bio-Based Products

In this part, we will discuss the three main products of the biological conversion method: bio-plastics, biofuels, and bio-alcohols. Producing these products and attempting to improve each process, as well as discovering useful new products, might serve as a road map for future research.

1.4.2.1 Bioplastics

Bioplastics are plastics derived in whole or in part from biological material. Bioplastics differ from biodegradable plastics, which are readily decomposed by microorganisms. Polyhydroxyalkoanates (PHAs) can be synthesized by microbes with the polymer accumulating in the microbes' cells during growth [55]. Packaging, food services, agriculture and horticulture, consumer electronics, and other industries are all using bioplastics. About 2.42 million tons of bioplastics were produced globally in 2021, with nearly 48% (1.15 million tons) of that volume going to the packaging market, which is the largest market for bioplastics (Fig. 1.8).[67].

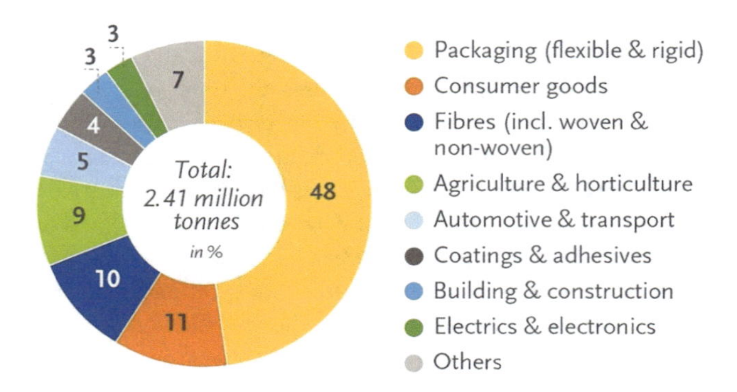

Fig. 1.8 Global production capacity of bioplastics in 2021 by marketing segment [67]

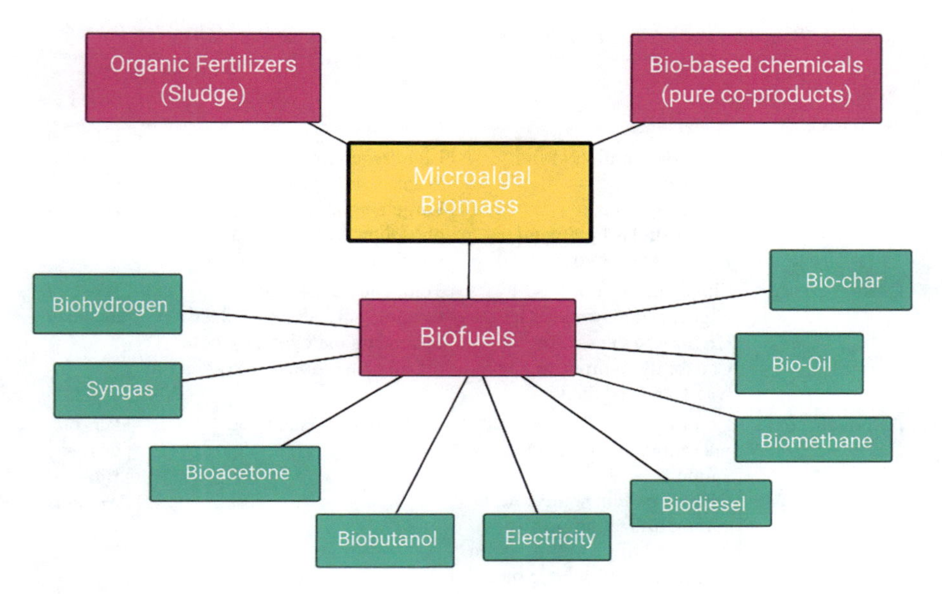

Fig. 1.9 Biofuel production from microalgae and two side products (Organic fertilizers, Bio-based chemicals). (Modified after [68])

1.4.2.2 Biofuels

Microalgae have been tested as a viable feedstock for biofuel generation in the current era due to its high energy content, rapid growth rate, low-cost culture methods, and significant ability for CO_2 fixation and O_2 addition to the environment. Biofuel has gained significant attention as an alternative fuel in recent years due to its capacity to adapt with gasoline for a maximum 85% blend without engine modification. As a result, academics and environmentalists are constantly questioning the suitability of various alternatives for biofuel. Figure 1.9 depicts the various forms of biofuels produced from microalgae; additionally, bio-based chemicals and bio-fertilizers are available as byproducts alongside biofuels [68].

1.4.2.3 Bio-Alcohols

Alcohols produced from biological resources or biomass are known as bio-alcohols. Bioethanol, the most common and extensively produced bio-alcohol, is an important alternative fuel for spark ignition engines. As ethanol has a poor energy density (70% that of gasoline) and is corrosive to current engine technology and fuel infrastructure, its use as a replacement for conventional gasoline is called into question. It also rapidly absorbs water, resulting in separation and dilution in the storage tank. Isopropanol can be produced biologically. It can be used to supplement

Table 1.2 Benefits, drawbacks, and products produced by microorganisms [58]

Organisms	Advantages	Disadvantages	Bio-based products
Algae	Wide distribution Fast growing Fast CO_2 uptake High cellular lipid content High-value byproducts	Light requirement Water requirement Large amount of phosphorousRequired as a fertilizer	Bio-plastics Biofuels
Cyanobacteria	Simple cultivation Higher photosynthetic levels Higher growth rates Capability to produce a wideRange of fuels	Temperature, pH, and lightIntensity affect productivity Increasing the operating cost ofCell cultivation due to agitation	Bio-alcohols
β-Proteobacteria	Aerobic microorganisms'Easier cultivation Diverse carbon sourcesAnd carbon utilization pathways Natural ability to store carbon Availability of genetic modification tools	Under development gas fermentation	Bio-plastics Bio-alcohols

gasoline. It is also used to esterify fat for biodiesel production instead of methanol, which lowers its tendency to crystallize at low temperatures [58, 69].

Table 1.2 lists the advantages, disadvantages, and products generated by all three microorganisms: algae, cyanobacteria, and proteobacteria.

1.5 Carbon Mineralization

Carbon mineralization is a natural process that occurs when CO_2 reacts with metal cations to generate carbonate minerals, with calcium and magnesium being the most attractive metals. The CO_2 is permanently eliminated from the atmosphere after being trapped in the permanent and nontoxic state of the carbonate minerals. Mineralization methods are generally classified into two types: in situ and ex situ. In situ mineralization or mineral trapping involves injecting CO_2 into geological formations containing alkaline minerals in order to promote natural carbon mineralization over time. Ex situ mineralization occurs when CO_2-bearing gases react with alkaline mine tailings or industrial wastes on the earth's surface in an industrial process. These approaches can also provide a low-cost way to reduce greenhouse gas emissions.

In general, the degree of mineral carbonation is determined by available CO_2 dissolved in solution, available alkalinity in solution, and chemical conditions that promote available alkalinity via mineral dissolution and carbonate precipitation [70]. Mineral carbonation products are stable solids that limit the possibility of

CO_2 emission back into the atmosphere. According to IPCC Special Report on Carbon dioxide Capture and Storage, the fraction of carbon dioxide stored by mineral carbonation retained after 1000 years in in situ mineralization is almost expected to be 100%. As a result, the need for monitoring disposal sites will be minimized [71]. Carbonation reactions that mineralize CO_2 are exothermic, so it does not require energy inputs, which means these spontaneous reactions generate heat. On the other side, mineralization processes happen very slowly and might take hundreds of years. This issue has to be resolved, especially with the ex situ mineralization approach, which calls for various energy-intensive pre-treatment procedures like grinding and heating [72, 73].

The mineralization potential capacity of resources due to the presence of appropriate geological formations and industrial wastes is virtually limitless. Ultramafic and mafic rocks like peridotite and basalt are more suited due to their high concentration of metals like magnesium and calcium compared to intermediate and felsic rocks like diorite and granite, which are made up of inert minerals like silicon dioxide. Basaltic rocks are the most feasible formation to store CO_2 as they make up most of the ocean floor, over 70% of the earth's surface, and more than 5% of the continents [73]. In addition, alkaline solid wastes such as iron/steel slags, coal-fired products, fuel combustion products, mineral processing wastes, incinerator residues, cement/concrete wastes, and pulp/paper mill wastes exist in Gt-Size for mineralized construction materials [74, 75]. Mineral carbonation technologies generally store between 10,000 and 1000,000 Gt of total carbon. In contrast, the estimated carbon production in 100 years is roughly 2300 Gt. Despite this enormous potential, large-scale carbon mineralization has not yet been implemented owing to the absence of information on mineral concentrations, compositions, and volumes at specific geologic resource locations [76, 77].

1.5.1 In-Situ Mineralization

The process of injecting CO_2 into geological formations containing alkaline minerals to enhance natural carbon mineralization over time is known as in situ mineralization or mineral trapping. In situ mineralization requires subsurface rocks rich in suitable alkaline minerals (magnesium and calcium), which can react with CO_2. Injection in gaseous, liquid, or supercritical forms into underground reservoirs is the three storage options for CO_2. In these systems, four types of trapping mechanisms are considerable for CO_2 utilization: *Hydrodynamic trapping* refers to CO_2 trapping as supercritical fluid or gas under a low-permeability caprock. *Residual trapping* refers to trapping CO_2 in tiny pores. *Solubility trapping* relates to the dissolution of CO_2 in the formation fluid. Finally, *mineral trapping* refers to the incorporation of CO_2 in a stable mineral phase via reactions with mineral and organic matter in the formation. As storage proceeds from structural to mineral trapping, CO_2 becomes more immobile, enhancing storage safety and lowering reliance on cap rock effectiveness (Fig. 1.10) [71, 73, 78].

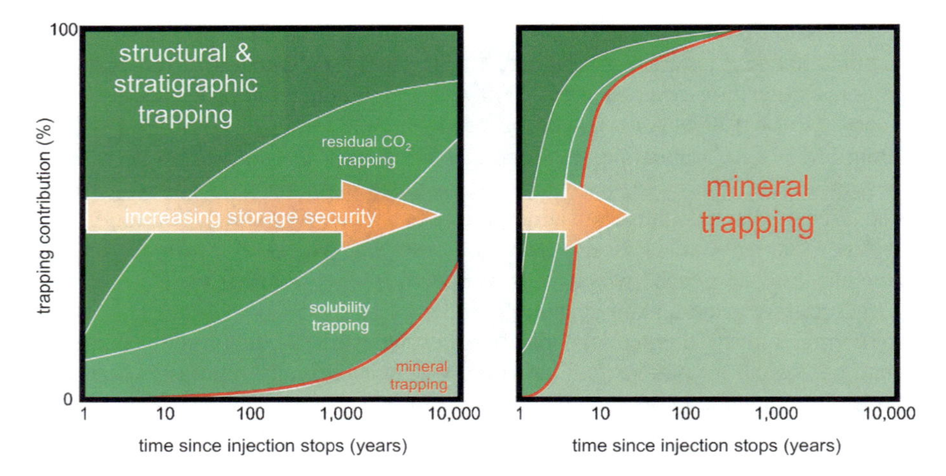

Fig. 1.10 Physical and geochemical trapping is used to ensure storage security. The physical process of residual CO_2 trapping and the geochemical processes of solubility trapping and mineral trapping, increase with time. The left-hand panel, typical sedimentary reservoir, right-hand panel, peridotite reservoir [70, 71]

Sedimentary basins are capable of implementing in situ mineralization. In these formations, the porosity and permeability of the target formation are essential factors in injectivity, while solution chemistry, temperature, and pH are crucial factors in carbonate formation potential [79]. However, this approach faces some significant problems. Low rock reactivity due to the lack of silicate-bound divalent metals required for carbonate production is the major challenge; the risk of returning CO_2 to the surface is also present, as the majority of the injected CO_2 will most likely remain in the gaseous, liquid, or supercritical phase for an extended period [71, 73]. As a result, several CCS approaches have been developed to overcome the constraints of sedimentary injection. The most important one is the injection of CO_2 into mafic or ultramafic lithologies that have large concentrations of divalent cations like Ca^{2+}, Mg^{2+}, and Fe^{2+} in order to promote fast mineralization to calcite ($CaCO_3$), dolomite ($CaMg(CO_3)_2$), magnesite ($MgCO_3$), or siderite ($FeCO_3$) [73, 80, 81]. Figure 1.11 depicts the mafic (basaltic), ultramafic, and sedimentary reservoirs accessible for carbon mineralization. Although mafic rocks are more plentiful in size, ultramafic rocks can react faster with CO_2 due to their more significant concentration of reactive minerals. Additionally, large-scale facilities and pilot projects for CO_2 sequestration across the globe are visible [72].

1.5.1.1 Challenges and Risks

Regardless of ex situ methods, in situ mineralization should be regularly monitored as it may confront some challenges and risks that must be addressed. Since direct sampling of mineralization is too complex and expensive, quick indirect monitoring

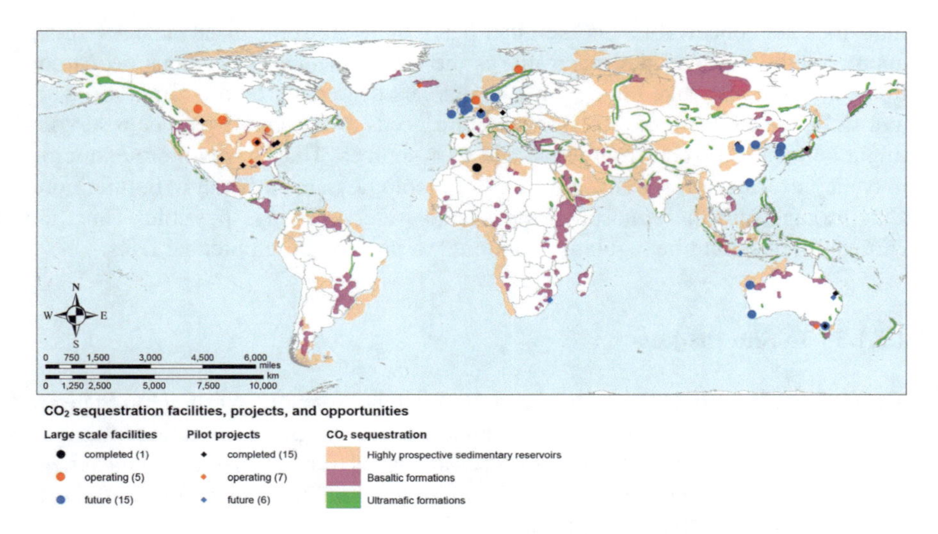

Fig. 1.11 Map of CO_2 sequestration facilities, pilot projects, and long-term storage potential in geological formations [72]

appears feasible and cost-effective. Leakage from wellbores or non-sealed fractures in the caprock and pressure buildup in the reservoir that may result in caprock hydraulic fracturing are significant risks [72]. Also, contamination of drinking water aquifers as supercritical CO_2 is buoyant in the subsurface and can travel upwards in the presence of an open pathway, such as a transmissive fault. Furthermore, injecting fluids underground can trigger earthquakes by increasing pore fluid pressure and changing rock volume, allowing faults to move [70, 76].

All these risks can be avoided by monitoring CO_2 plume migration, pressure in and above the reservoir, induced seismicity, the degree of secondary trapping mechanisms, leakage into groundwater, and the chemistry of freshwater aquifers near the CO_2 reservoir and leakage to the atmosphere. In terms of human health, utilizing best practices and managing operations to reduce the likelihood of worker injury, uncontrolled CO_2 emissions, and fugitive emissions are also crucial [70, 72].

1.5.1.2 Pros and Cons

Compared to ex situ mineralization, in situ mineralization has several benefits. The first and most important are the readily available, vast rock "reservoirs" that may be used to absorb CO_2 and reduce its effects on the environment. These reservoirs may also be found all over the globe, as seen in Fig. 1.11. This approach is more advantageous regarding costs and energy since, despite ex situ mineralization, no pre-treatment activities are required. Finally, because of the large-scale projects that may be performed using this approach, the foundation of government and big industries are conceivable. On the other hand, there are some disadvantages to this approach: the first and the major one is the slow kinetics of reactions, as carbon mineralization may take up to hundreds of years depending on the formation types and CO_2 injection. Moreover, infrastructure needs are prohibitive since reservoirs might be located distant from waste and CO_2 sources. That is why more extraordinary engineering efforts and advanced technologies are necessary. Furthermore, CO_2 leakage into the atmosphere or ground water is always possible. Thus, the entire system should be regularly monitored to prevent these potential risks.

1.5.1.3 In Situ Projects

The CarbFix experiment in Iceland and the Wallula Project in Washington State are the two projects that have shown in situ mineralization of CO_2 in basaltic formations. In both experiments, thick sequences of basaltic lavas were extensively characterized regarding composition, structure, and hydrology before injecting CO_2-rich fluids to test storage in pore space and produce solid carbonate minerals.

1.5.1.3.1 CarbFix

The CarbFix Pilot Project is an academic-industrial collaboration that has created an innovative method for safely and permanently capturing CO_2 and H_2S from emission sources and storing it as stable carbonate minerals in the subsurface basalts by imitating and speeding up the natural process of carbon mineralization. With this method, CO_2 and other acid gases may be captured and stored as stable mineral phases for less than $25 per ton [82]. It involves a combined program consisting of a CO_2 pilot gas separation plant, CO_2 injection pilot test, laboratory-based experiments, studying of natural analogs, and numerical modeling. Following CO_2 injection into aquifers, it will dissolve and acidify the formation water before dissociating into bicarbonate and carbonate ions via the following reaction (Eq. 1.31) [83]:

$$CO_2(aq) + H_2O \leftrightarrow H_2CO_3 \leftrightarrow HCO_3^- + H^+ \leftrightarrow CO_3^{2-} + 2H^+ \quad (1.31)$$

The subsurface injection of carbonated water causes it to react with the Ca and Mg found in the rock. Rocks often include calcium and magnesium as oxides. However, since many rocks, like basalt, include silicate minerals of these elements (like forsterite and anorthite), some example reactions may be as follows (Eqs. 1.32 through 1.34) [83, 84]:

$$(Ca, Mg)^{2+} + C_2O + H_2O \rightarrow (Ca, Mg)CO_3 + 2H^+ \quad (1.32)$$

$$Mg_2SiO_4 + 4H^+ \rightarrow 2Mg^{2+} + 2H_2O + SiO_2(aq) \quad (1.33)$$

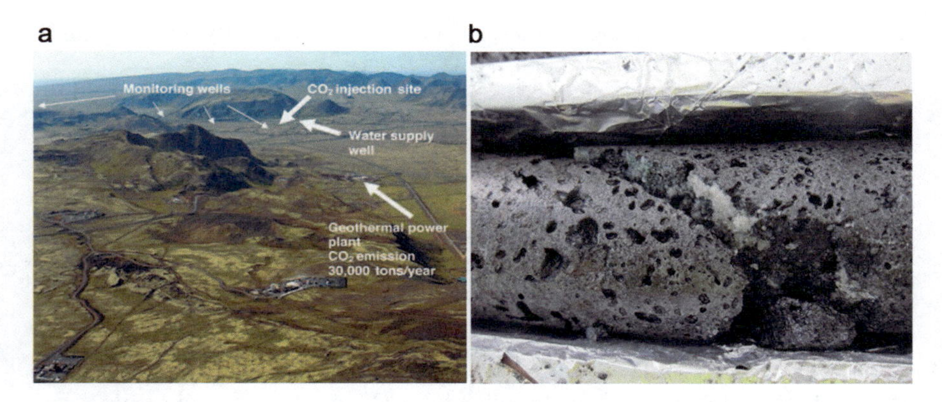

Fig. 1.12 (**a**) The field-scale, in situ basalt-carbonation pilot plant in Hellisheidi, Iceland [84], (**b**) Core from CarbFix site. (Source: CarbFix project, Orkuveita Reykjavikur)

$$CaAl_2Si_2O_8 + 2H^+ + H_2O \rightarrow Ca^{2+} + Al_2Si_2O_5(OH)_4 \tag{1.34}$$

The CO_2 gas injection site is located in southwest Iceland, about 3 km south of the Hellisheidi geothermal power plant above subsurface basalts formations. (Fig. 1.12) The power plant has a CO_2 generation capability of around 60,000 tons per year. A treatment facility separates the primary gases generated, which include CO_2 and H_2S. The H_2S is separated and injected back into the geothermal reservoir, while the CO_2 (98% CO_2, 2% H_2S) is transported through a 3 km long pipeline to the CO_2 injection location. The CO_2 injected into the storage formation entirely dissolves in water, resulting in a single fluid phase entering the storage formation. CO_2 at 25 bar and groundwater are injected together. Carbon dioxide is transported to a depth of 500 meters by injected groundwater, where it enters the target storage formation totally dissolved. Under these circumstances, CO_2-charged water reacts with basaltic minerals, increasing pH and alkalinity. Given that the amount of water necessary to completely dissolve CO_2 varies on the temperature and partial pressure of CO_2, the total dissolution of CO_2 at the CarbFix site takes 22 tons of H_2O per ton of CO_2 [83].

By utilizing tracers such as trifluormethylsulfur pentafluoride (SF_5CF_3), acid red dye (amidorhodamine G), and radiocarbon (^{14}C), the mineralization of the injected gases has been demonstrated and is being tracked by sampling fluids from wells close to the injection spot. The injection well is filled with known quantities of CO_2 and tracers. The assessment of CO_2 mineralization by mass balance calculations is made possible by measured tracer concentration and chemical composition in monitoring wells. Utilizing various isotopes, the mineralization has also been quantified. According to monitoring results, more than 95% of the subsurface CO_2 injections mineralized within a year, and almost all of the H_2S injections mineralized within 4 months after injection. Furthermore, the injected radioactive carbon tracer was found in the carbonates that precipitated on the pump and inside of the monitoring well pipes. This finding demonstrated that carbon dioxide may be quickly and permanently trapped in basaltic bedrock, consequently lowering greenhouse gas emissions [85].

The new project CarbFix2 builds upon the success of the original CarbFix project, which was funded by the EU's seventh Framework Program. It is a comprehensive project consisting of [86]:

- Development of the technology to perform the CarbFix geological carbon storage method using seawater injection into submarine rocks
- Reducing the cost of the entire CCS chain
- Impure CO_2 capture and co-injection into the subsurface
- Integration of the CarbFix method with novel direct air capture technology

The goal of the CarbFix2 project was to make the CarbFix geological storage solution both commercially feasible with a full CCS chain and transportable across Europe.

1.5.1.3.2 Wallula Project

The Wallula Project in Washington State, the world's first continental flood basalt sequestration, was conducted in 2013 by the Pacific Northwest National Laboratory (PNNL) of the U.S. Department of Energy Big Sky Regional Carbon Sequestration Partnership to examine the viability of safely and permanently storing CO_2 in basalt formations. By injecting 1000 metric tons of supercritical CO_2 into a natural basalt formation in the Columbia River Basalt Group at 830–890 m depth, PNNL researchers started a field demonstration of carbon storage. Prior to drilling, site appropriateness was evaluated by collecting, processing, and analyzing a four-mile, five-line, three-component seismic swath that was processed as a single data-dense line. Results from 2 years of post-injection monitoring, including a long-term sampling of water retrieved from the injection zone, shallow groundwater and soil gas monitoring, and PSInSAR, [87] revealed the formation of new carbonate minerals as a result of CO_2 injection. Nodules of calcium, iron, magnesium, and manganese carbonate mineral ankerite ($Ca(Fe, Mg, Mn)(CO_3)_2$) were detected in vesicles throughout the cores. Additional carbon isotope research confirmed the nodules to be chemically unique from basalt's naturally occurring carbonates and to be in direct accordance with the isotopic signature of injected CO_2. At the top of the injection zone, there was unmineralized CO_2 that was still present beneath the caprock, showing that not all of the CO_2 had mineralized (Fig. 1.13). Results from modeling show that within 2 years, mineralization sequestered almost 60% of the CO_2 that was injected. However, it is uncertain what will happen to the remaining CO_2 because no leaks have been identified. According to the experimental results, carbonates only occupied around 4% of the reservoir accessible pore space, giving it a significant amount of storage capacity [76, 88].

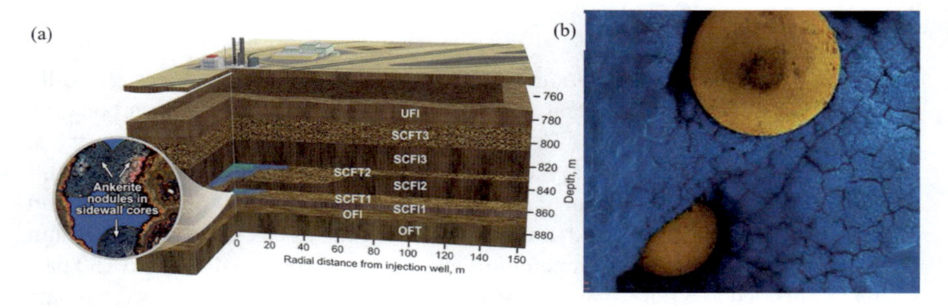

Fig. 1.13 (**a**) Schematic representation of the Wallula Project and location of Ankerite nodules forming in the deep subsurface and (**b**) calcium carbonate nodules. (Source: Odeta Qafoku I PNNL)

1.5.2 *Ex Situ Mineralization*

Ex situ mineralization takes place when CO_2-bearing gases in an industrial process interact with natural minerals, alkaline mine tailings, or industrial wastes on the earth's surface. The source material, which is frequently an alkaline earth metal silicate, is transformed into the metal's carbonate as a result of this reaction. The finished product, depending on the type of raw material, may be valuable and utilized as feed in downstream industries. One of the most significant advantages of this technology is waste management, which results in the production of a valuable product with fewer environmental problems after the reaction of hazardous wastes from industries such as iron and cement, which can damage water, soil, and even atmosphere. The proximity of some of these raw materials to point sources of CO_2, the size of the available tailings, which eliminates the need for energy-intensive processes, and the faster reaction time are some additional benefits of this method over the in situ method, in addition to the cases already mentioned. On the other hand, there are significant obstacles to this technology that must be addressed as quickly as feasible, such as the high cost per kilo of carbon captured when compared to the in situ method. Furthermore, while some carbon mineralization products have commercial value, the low value of other production materials is not yet convincing to invest in this technology, and as a result, despite the enormous potential of the raw materials, employing this approach on a large scale is not common in the globe.

1.5.2.1 Ex Situ Sequestration Routes

Carbonation studies have identified several ways for performing ex situ CO_2 sequestration, which are classed as direct carbonation and indirect carbonation. Each of these two approaches will be discussed more below.

1.5.2.1.1 Direct Carbonation

The process of direct carbonation is separated into two parts: direct gas-solid carbonation and aqueous mineral carbonation. The direct gas-solid carbonation process is the simplest method. The potential of this method for heat recovery at high temperatures reduces energy consumption and improves viability. Unfortunately, this approach has fundamental difficulties, including a slow reaction rate, and is applicable only for refined and unusual materials such as calcium and magnesium oxides and hydroxides. High temperatures and pressures (between 100 and 150 bar) are recommended as a remedy to this issue, although this approach may decrease the process overall efficiency due to the significant amount of energy needed. The direct gas-solid reaction of olivine serves as an illustration of this process (Eq. 1.35) [71, 89].

$$Mg_2SiO_4(s) + 2CO_2(g) \rightarrow 2MgCO_3(s) + SiO_2(s) \tag{1.35}$$

On the other hand, aqueous mineral carbonation is the most commonly studied ex situ mineral carbonation route, and it was one of the first that was investigated on a small scale [90]. The carbonic acid pathway technique comprises CO_2 interacting with olivine or serpentine in an aqueous solution at high pressure (100–159 bar). This process involves dissolving CO_2 in water, where it dissolves into bicarbonate and H^+, producing a pH of around 5.0 to 5.5 at high CO_2 pressure. If we use the previous aqueous carbonation process as an example, the reactions are as follows (Eqs. (1.36) though 1.38) [89]:

$$CO_2(g) + H_2O(l) \rightarrow H_2CO_3(aq) \rightarrow H^+(aq) + HCO_3^-(aq) \tag{1.36}$$

$$Mg_2SiO_4(s) + 4H^+(aq) \rightarrow 2Mg^{2+}(aq) + SiO_2(s) + 2H_2O(l) \tag{1.37}$$

$$Mg^{2+}(aq) + HCO_3^-(aq) \rightarrow MgCO_3(s) + H^+(aq) \tag{1.38}$$

Mg^{2+} is released by H^+ in the second reaction, and in the third reaction, it reacts with bicarbonate to form magnesium carbonate, which subsequently precipitates. As with the prior method, raising the temperature and pressure can enhance the reaction rate. Furthermore, pre-treatment methods such as crushing and heating can be used to improve carbonate conversions and acceptable reaction rates; however, it should be noted that the use of these techniques, despite improving the process, increases energy consumption, resulting in a reduction in stored carbon [91].

1.5.2.1.2 Indirect Carbonation

Since direct methods for unrefined solid materials are ineffective, there is a strong need for alternative methods like indirect mineral carbonation that are more energy efficient and cost-effective acids or other solvents are used in this multi-stage process to extract reactive components from minerals. The extracted components then react with CO_2 in either an aqueous or a gaseous phase. Indirect carbonation, like direct methods, can be divided into some categories.

The first method that we discuss here is direct gas-solid carbonation. In order to improve the conversion rate, the mineral could first be converted into an oxide or hydroxide and subsequently carbonated. The direct gas-solid carbonation of calcium/magnesium oxides/hydroxides proceeds much faster than the gas-solid carbonation of calcium/magnesium silicates, although a high temperature and CO_2 pressure are required. As a result, in the first stage of this method, which typically occurs in a fluidized bed, alkaline earth metals in the silicate form are changed into oxide or hydroxide form. Following this reaction with CO_2, the products of this step react with CO_2 and precipitate as stable carbonates (Eqs. 1.39 through 1.41) [92]:

$$Mg_2SiO_4(s) + 4HCl(g) \rightarrow 2MgCl_2(aq) + 2H_2O(l) + SiO_2(s) \tag{1.39}$$

$$MgCl_2(aq) + 2H_2O(l) \rightarrow Mg(OH)_2(s) + 2HCl(aq) \tag{1.40}$$

$$Mg(OH)_2(s) + CO_2(g) \rightarrow MgCO_3(s) + H_2O(l) \tag{1.41}$$

In addition to the procedure mentioned above, using various acids such as acetic acid and hydrochloric acid is also frequent. The goal of applying these acids is to maximize Ca and Mg ion leaching while ensuring selective leaching. Because acetic acid is more acidic than ammonium chloride, it has a higher calcium ion leaching ratio [93]. The use of acetic acid as an extractant has a major side effect of lowering the pH of the leachate. Alkali must be used to stimulate the carbonation reaction in order to fix this problem [94]. As seen in the reactions below (Eqs. 1.42 and 1.43), divalent magnesium is separated in the first stage of the magnesium silicate reaction with acetic acid and is then ready to react with carbon dioxide gas in the next stage:

$$MgSiO_3(s) + 2CH_3COOH(aq) \rightarrow Mg^{2+}(aq) + 2CH_3COO^-(aq) + SiO_2(s) + H_2O(l) \tag{1.42}$$

$$Mg^{2+}(aq) + 2CH_3COO^-(aq) + H_2O(l) + CO_2(g) \rightarrow MgCO_3(s) + 2CH_3COOH(aq) \tag{1.43}$$

Ammonium chloride is a kind of strong acid and weak alkali salt. For the leaching reaction using ammonium chloride, the solution shows alkalinity as the reaction proceeds because of the generation of ammonia. Noteworthy, the alkalinity of the solution promotes the dissolution of CO2 in the precipitation reaction. At the same time, the leachate using ammonium chloride has a strong pH-buffer ability, because an ammonia buffer solution is formed in it. Ammonium chloride is regarded as an ideal recyclable solvent because it can be regenerated in the carbonation reaction stage. As the carbonation reaction proceeds, NH4Cl is regenerated, which makes it recyclable for the leaching reaction [93]. As shown in the reaction below (Eqs. 1.44 and 1.45), in addition to the formation of magnesium carbonate at the end of the reaction, ammonium chloride is also generated, saving the consumption of this acid throughout the cycle:

$$2MgSiO_3(s) + 4NH_4Cl(aq) \rightarrow 2MgCl_2(aq) + 4NH_3(g) + 2H_2O(l) + SiO_2(s) \quad (1.44)$$

$$2MgCl_2(aq) + 4NH_3(g) + 2CO_2(g) + 2H_2O(l) \rightarrow 2MgCO_3(s) + 4NH_4Cl(aq) \quad (1.45)$$

Other solvents commonly used in indirect carbonation include ammonium sulfate, citric acid, hydrochloric acid, sulfuric acid, and others. In relation to the use of solvent, it is important to note that, despite improvements in the reaction rate and overall efficiency, if these materials are not recovered, there is a risk of serious environmental damage, particularly to the local ground water and soil, so all aspects of using these materials must be considered.

1.5.2.2 Feedstocks

The feedstocks needed for the ex situ reaction with CO_2, depending on where they come from, can be categorized into three main groups: natural minerals, mine tailings, and industrial waste. The three cases are further discussed in the following sections.

1.5.2.2.1 Natural Minerals

Natural minerals such as wollastonite ($CaSiO_3$) and forsterite (Mg_2SiO_4) are considered suitable for mineralization owing to the presence of alkaline earth elements such as Ca and Mg. Although alkali metals like Na and K have the capacity to react with CO_2 and capture it, they are less frequently utilized as an efficient raw material due to the strong reactivity of their final product, particularly in water. Additionally, iron can be a useful source due to its abundance in the ground and its great capacity to react with CO_2 and produce siderite ($FeCO_3$), but its usage is not cost-effective due to the high value of metal. Natural minerals suited for CO_2 reactions are classified into two types: natural calcium silicates such as wollastonite ($CaSiO_3$) and natural magnesium silicates such as olivine (Mg_2SiO_4) and serpentine ($Mg_3Si_2O_5(OH)_4$). Compared to magnesium silicate, minerals in the first category – natural calcium minerals like wollastonite – have a quicker reaction rate and a wider range of industrial applications. However, the widespread availability of magnesium silicates in a variety of forms, including dunites, serpentinites, and peridotites, has made them a dependable source for producing stable carbonates [95]. The most important reactions between natural minerals and CO_2 that result in stable carbonate are shown below (Eqs. 1.46 to 1.48):

$$\text{Wollastonite}: CaSiO_3 + CO_2 \rightarrow CaCO_3 + SiO_2 \quad (1.46)$$

$$\text{Olivine}: Mg_2SiO_4 + 2CO_2 \rightarrow 2MgCO_3 + SiO_2 \quad (1.47)$$

$$\text{Serpentine}: Mg_3Si_2O_5(OH)_4 + 3CO_2 \rightarrow 3MgCO_3 + 2SiO_2 + 2H_2O \quad (1.48)$$

One of the most significant benefits of employing natural minerals for carbon mineralization is the abundant availability of these materials on a huge scale when compared to alternative sources such as industrial wastes. However, the unprocessed nature of these materials and the requirement for pre-treatment procedures like grinding and crushing to create an effective surface area are some important drawbacks of this approach. Furthermore, the necessity for transportation due to the sources' considerable distance from CO_2 point sources raises the price and lessens the appeal of this strategy.

1.5.2.2.2 Mine Tailings

Mine tailings are the byproducts of mineral processing operations. These tailings are a slurry of pulverized rock, as well as water and chemical reagents left over after processing. Their phase and chemical compositions vary depending on the characteristics of source rocks and the mineral processing procedures they have experienced. Mining tailings have always been seen as having little or no financial value. But the utilization of mining tailings has advanced to a new level as a result of recent technological advancements and new demands that have emerged across many industries. This new approach has been made most appealing by the reactivity and alkalinity of mineral tailings, which has found application in processes like acid neutralization (for example, use in reducing the environmental effects of acid mine drainage), reducing carbon in the atmosphere (as one of the environmental priorities of the twenty-first century) and long-term immobilization of environmentally hazardous metal. Due to the presence of reactive elements like Ca and Mg, the utilization of ultramafic mineral tailings offers the possibility of eliminating millions of tons of CO_2. In addition, the large amount of reactive surface area observed in crushed tailings is appropriate for reacting with CO_2. This eliminates the need for an energy-intensive operation such as crushing (compared to the use of natural and raw minerals).

The likelihood of getting these tailings will grow day by day as a result of the rise in demand in the mining industry in the upcoming years. This makes things simpler for heavy industry companies to employ these materials to decrease environmental pollution, especially in order to meet zero-carbon targets. However, there are also significant issues that require adequate attention, such as energy-intensive pre-treatments like heat treatment, and chemical activation with reagents. Furthermore, due to the placement of mine tailings in remote areas, one of the limits that challenge the use of this technology is the necessity to transport them.

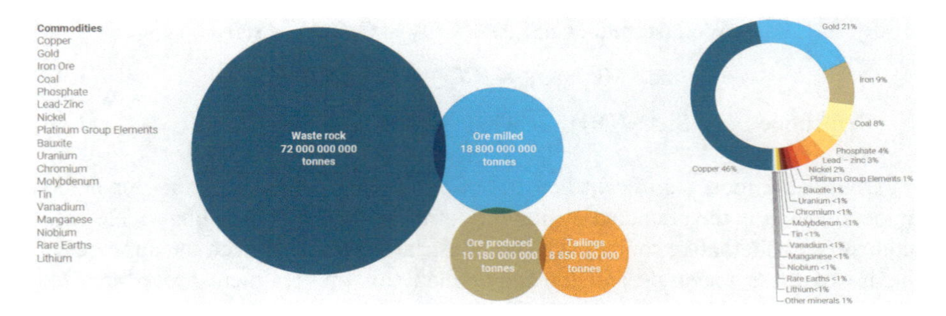

Fig. 1.14 Estimate of tailings and waste rock produced in relation to ore production and worldwide proportion of tailings per commodity in 2016 [96]

Each year, the amount of tailings generated, particularly in open pit mines, increases significantly due to a drop in the grade of extractable rocks. Only in 2016, nearly 9 billion tons of tailings from metal and mineral extraction were generated, creating challenges, especially in the field of maintenance and prevention of harmful environmental effects. It should be emphasized that copper, gold, iron, and coal accounted for the majority of this tailings, with 46, 21, 9, and 8%, respectively (Fig. 1.14) [96].

Nickel and asbestos are the primary sources of ultramafic tailings. Manufacturing sites can be used in carbon sequestration of each of these tailings, which is a combination of much unique magnesium and calcium-containing compounds, and their dissolving rate and reactivity are related to their composition. As a result of these characteristics, four unique patterns of CO_2 reactivity in ultramafic tailings may be imagined [76]:

- Fast carbonation of the magnesium hydroxide mineral like brucite
- Fast absorption of CO_2 by hydrotalcite minerals
- Fast cation exchange reactions of swelling clays
- Relatively slow dissolution of calcium and magnesium silicate

Nickel Tailings

Nickel is mined from two different types of deposits: nickel-rich laterite generated by weathering of ultramafic rocks in tropical regions containing garnierite (Ni-silicate) and from Ni-sulfide concentrations in mafic igneous rocks, primarily pentlandite. Despite the high costs of employing nickel tailings, because of the high MgO content, it is possible to integrate extraction and CO_2 separation using innovative methods. Furthermore, ultramafic deposits of nickel support stabilization of chrysolite asbestos and decrease the environmental impact of these tailings [95]. In 2011, the world's nickel resources were projected to be 296 million tons (Fig. 1.15). This quantity is divided into 178 million tons for nickel laterite deposits and 118 million tons for nickel sulfide resources. Australia has the most considerable

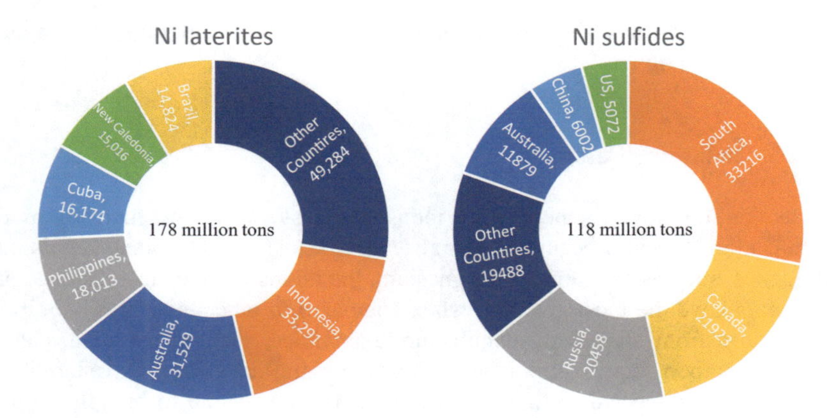

Fig. 1.15 Laterite and sulfide nickel deposits in several countries in 2011(numbers are in KT). (Modified after [97])

nickel resources than any other country, with 31 million tons of laterite resources and 11 million tons of sulfide resources. Indonesia and South Africa are in the next places with reserves of 33 million tons [97]. The abundance of nickel deposits and their distribution across continents allow this material to be employed as one of the essential resources in lowering existing carbon and reaching zero-carbon technologies in related sectors. There are difficulties in extracting nickel from low-grade ultramafic deposits. Serpentine minerals are typically found in ultramafic ores. These ores have low recoveries because of the difficulty in dispersing and effectively rejecting them. For instance, during the first five years of operation at Mt. Keith, Australia, nickel recovery from ores containing 0.58% Ni and 40% MgO was only 60% [91].

Asbestos Tailings

Asbestos is a naturally occurring category of fibrous materials. There are six types of asbestos that have been discovered; they come from the amphibole and serpentine mineral groups. White asbestos, often known as chrysotile ($Mg_3(Si_2O_5)(OH)_4$), is the kind of asbestos that is most frequently found in veins in serpentine rock formations. Where serpentine is mined for chrysotile asbestos, the tailings often include considerable residual asbestos and may be categorized as hazardous. These tailings would be great feed for mineral carbonation because not only has size reduction occurred, but when chrysotile is carbonated, the asbestiform character of minerals is removed and it is highly environmentally beneficial as asbestos can cause cancer of the lung, cancer of the larynx, and certain gastrointestinal cancers. Globally, 4 Mt. of asbestos is produced, each ton producing 20 tons of tailings. Because of the high quantities of MgO (40%) found in these tailings, they would constitute an excellent source of mineral carbonation [95, 98]. Despite the benefits that may be obtained from the carbonization of asbestos, the world's extraction of this material is significantly declining owing to its environmental concerns, making it impossible to utilize asbestos as a viable feedstock for carbonization and mitigating global warming in the long term.

1.5.2.2.3 Industrial Wastes

This section investigates the use of industrial waste as a raw material in the mineralization process. Because of the existence of considerable amounts of alkaline earth metals, such as calcium and magnesium, the tailings of the steel, cement, and coal sectors have the most potential when compared to other industries. Additionally, residues from aluminum manufacturing facilities, such as red mud, can be utilized for carbon sequestration. Because of the rising need for the availability of more products connected to these industries, it is conceivable to broadly employ these raw materials to reduce environmental consequences. The fact that industrial wastes, as opposed to mineral tailings, are situated close to point sources of CO_2 emission, decreases the cost of the process and also improves the likelihood that these products will react and create stable carbonate minerals. As a consequence, in addition to capturing carbon from the atmosphere, the approach has been proposed to manage unstable industrial wastes for disposal in compliance with safety regulations, as well as their reuse.

Steel Slag

Steel slag is a waste product produced during the manufacturing of steel. It is massively produced during the steelmaking process utilizing electric arc furnaces. Steel slag can be produced when iron ore is smelted in a basic oxygen furnace. These slags are mostly used as aggregate replacement in construction applications such as granular foundations, embankments, engineered fill, highway shoulders, and hot mix asphalt pavement.

Steel slags are generally classified into four types: blast furnace slag (BF), basic oxygen furnace slag (BOF), electric arc furnace slag (EAF), and ladle furnace slag (LF). Table 1.3 shows the most common components of these four categories. CaO, MgO, Al_2O_3, SiO_2, and Fe_2O_3 are the basic chemical compositions of slag. The chemical compositions of different slags vary substantially; CaO % in BF and LF slag is the highest, followed by BOF and EAF slag. Each slag has a roughly equal

Table 1.3 Most common chemical compositions of four slag categories [93]

Components slag type	CaO	MgO	Al_2O_3	SiO_2	Fe_2O_3	TiO_2	MnO	Cr_2O_3	Others
BF slag	42.67	8.57	13.21	29.41	0.37	1.49	0.40	0.001	3.879
BOF slag	42.43	9.15	3.03	12.00	26.74	0.48	2.85	0.22	3.10
EAF slag	32.30	5.01	2.74	28.83	23.53	1.06	2.40	0.11	4.02
LF slag	50.50	11.90	18.60	12.90	1.60	–	–	–	4.50

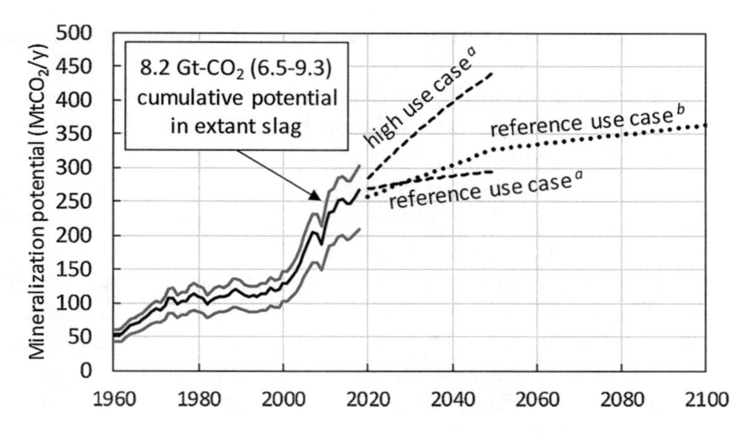

Fig. 1.16 The emission reduction potential of legacy and future iron and steelmaking slag by way of CO_2 mineralization [99]

MgO concentration. BF slag includes more SiO_2 and Al_2O_3, but BOF and EAF slag have more Fe_2O_3 [93].

Steelmaking activities emit considerable amounts of CO_2 (6–7% of total CO_2 emissions globally; 0.28–1 ton of CO_2/ton of steel produced). In addition, 315–420 Mt./y of slag is generated annually, according to estimates, although specific slag production numbers are not available [91]. Currently, slag-based CO_2 mineralization has the potential to cut emissions by 268 Mt. CO_2/y. Legacy slag has an 8.2 $GtCO_2$ mineralization potential, despite being frequently bonded in building material (Fig. 1.16) [99].

Although steel slag has been employed in various industrial-scale applications, there are still limitations associated with this technology. The most pressing issues that must be addressed are a lack of steel slag due to their widespread use in other industries, an increase in energy and economic costs while optimizing process parameters, limitations of reaction kinetics, minimizing environmental impacts, and a drastic difference in compositions for each waste unit, which makes it impossible to use a particular method on a global scale [76, 100].

Red Mud

Red mud, usually referred to as bauxite residue, is a byproduct of the Bayer process, which extracts alumina from bauxite ore. It is composed of a mixture of solid and metallic oxides and contains compounds like Fe_2O_3, Al_2O_3, TiO_2, CaO, SiO_2, and Na_2O. Annually, 70 million tons of red mud is generated, 1.0–1.5 t for each ton of alumina produced [91]. Red mud includes toxic heavy metals, and its high alkalinity makes it exceedingly corrosive and harmful to soil, water, land, air, and living forms, posing a significant disposal challenge. Although around 4 million tons of red clay is employed annually in the cement, iron, and road construction sector, this amount remains relatively small in comparison to the enormous rate of production.

Therefore, attempts to discover new applications for this hazardous waste must be continued.

Red mud can hold up to 0.01% of CO_2 emissions from fossil fuels globally, assuming they have a 5% CO_2 uptake. This equals to 3.5 Mt. of CO_2 every year. This amount of red mud created has the potential to prevent up to 0.01% of worldwide CO_2 emissions caused by fossil fuels [91, 95]. Various methods have been used for the neutralization of red mud by adding liquid carbon dioxide, saline brines or seawater, Ca and Mg-rich brines, soluble Ca and Mg salts, acidic water from mine tailings, fly ash, and carbon dioxide gas [101]. Despite all the benefits of adopting it, there are several issues that must be addressed in its deployment. The most significant issue in applying this technology on a large scale is the development of used devices with high capacity and low energy costs. The usage of this approach may assist in mitigating climate change effects and reduce the environmental problems associated with wastes if the aforementioned issues are resolved.

Coal Ash

Coal ash, also known as coal combustion residuals or CCRs, is largely created by the combustion of coal in coal-fired power plants. This ash contains a number of byproducts produced from the burning of coal, including fly ash, bottom ash, boiler slag, and clinker. When fine coal is burnt, a fine, powdery silica substance known as fly ash is produced. Bottom ash, on the other hand, is a larger coarse ash particle that accumulates at the bottom of a coal furnace because it is too big to be removed by smokestacks. Fly ash and bottom ash make up the majority of coal ash, making up 85–95 weight percent and 5–15 weight percent of all generated ash, respectively [95]. India, China, and the United States are now the greatest producers of fly ash, whereas nations such as the Netherlands, Italy, and Denmark have the highest utilization rates of produced coal fly ash (CFA) (Fig. 1.17) [102].

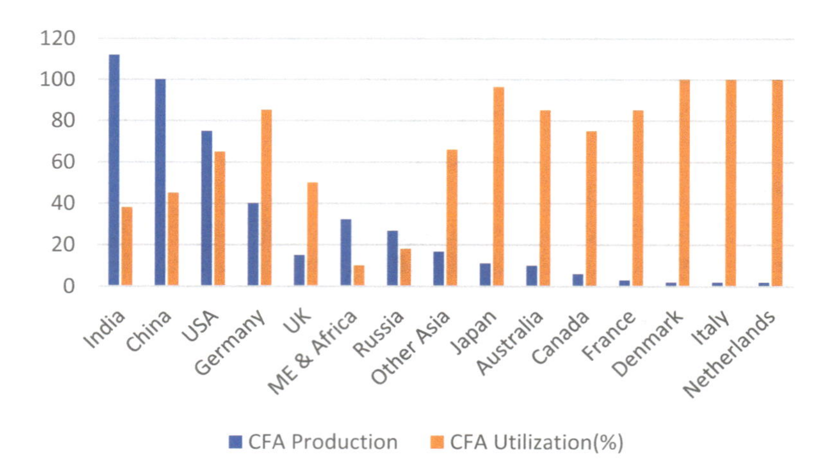

Fig. 1.17 Production and utilization of CFA across the globe. (Modified after [102])

Fly ash is applied in a variety of fields, including construction, as a cheap adsorbent for the removal of organic compounds, flue gas, and metals, lightweight aggregate, mine backfill, road sub-base, and zeolite synthesis, which is on the agenda to address environmental issues associated to fly ash [103, 104].

One of the primary benefits of employing fly ash for carbonation is the absence of pre-treatment activities, which are energy-intensive and can destabilize the entire process. This lack of necessity is due to highly fine granulation, which gives a high amount of material for reaction with CO_2 gas. Despite this significant property, the relatively low quantity of alkaline earth metals such as calcium and magnesium in these tailings limits their ability to be used for carbonation on a wide scale and at a cheap cost. This makes fly ash with a high lime concentration one of the most desirable raw materials for mineralization. These carbonation processes produce cement solids, which may be utilized to manufacture concrete. China is one of the most significant producers of these raw materials in the world, producing 100 million tons of fly ash each year, around half of that is used as a raw material for processes in other industries. However, as the country's rate of construction declines, there will soon be less demand for fly ash in concrete and paving, highlighting the need to find new applications for the material. In addition to aiding in the capture of carbon dioxide that has been released into the atmosphere, carbonation may be used in this situation to convert fly ash from a serious environmental threat into a less hazardous substance. These environmental effects include the accumulation of heavy metals like lead and arsenic, as well as ash particles in the air, which decrease air quality and expose people to these poisons through inhalation [76].

Cement

Cement – a fine powdered substance – is the most significant building material. It is a binding agent that sets and hardens to keep building components like stones, bricks, and tiles together. It is made mostly of limestone, sand or clay, bauxite, and iron ore; however, it can also contain other materials including shells, chalk, marl, shale, clay, blast furnace slag, and slate. There are different types of cement for different construction works and ordinary Portland cement (OPC) is the most commonly used type of cement in the world. Annual global cement production is 2.8 GT, with a projected growth to 4.0 GT in near future [91]. Cement manufacturing is the energy and carbon-intensive industry. The cement industry contributes approximately 5% of the global man-made carbon dioxide (CO_2) emissions and is thus becoming the second largest CO_2 contributor in the industry after power plants [105].

Numerous strategies have been suggested and put into practice to minimize the carbon emissions associated with the production of enormous quantity of cement globally and the constantly rising demand for this essential commodity. The utilization of supplementary cementitious materials, electric or hydrogen-fired kilns, point source carbon capture during cement manufacturing, and carbon mineralization are all examples. If these strategies are extensively implemented, the idea of attaining a carbon-neutral program for this industry is not far-fetched. The employment of three techniques – mixing carbonation (injecting pure CO_2 during concrete mixing), carbonation curing (changing water or steam with pure CO_2 during processing), and the creation of synthetic aggregates (reaction of CO_2 with alkaline feedstock containing calcium and/or magnesium, including recycled concrete and a variety of industrial wastes) – is more effective when it comes to the strategy of carbon mineralization for cement [76]. By increasing the strength of concrete during production, carbonation can reduce the amount of cement needed overall, reducing carbon intensity and feedstock costs.

The cement industry also produces a significant amount of wastes, such as cement kiln dust (CKD) and cement bypass dust (CBD). In fact, for every 100 tons of cement, 15–20 tons of CKD is produced. Cement waste is very reactive due to its fine particle size and high CaO content (20-60%). CKD generally contains 38–48% CaO, but because it also contains 46–57% $CaCO_3$, a substantial portion of it is already carbonated. CBD, on the other hand, has fewer carbonates than CKD. As a result, they have a high inclination to store CO_2 (0.5 ton CO_2 per ton CBD) [95]. Many factors, including the significant amount of usable raw materials, the simplicity of using raw materials due to the absence of energy-intensive pretreatment processes like crushing, and the high potential for CO_2 reaction, have led researchers to consider the uptake and mineralization of carbon by cement wastes.

1.5.2.3 Application and Products

The final products of mineral carbonation are numerous and can be utilized in various fields; our goal in this section is to review these products and their uses in various industries. The construction industry uses most the application of silica and carbonate materials, whereas cement and the resulting material, i.e., concrete, are manufactured on a Gt scale globally every year, and a substantial portion of the energy and carbon emitted into the atmosphere is the result of this massive volume of manufacturing. Other important environmentally friendly applications of these materials include use as materials in the process of mine rehabilitation and use as materials to reduce water and soil pollution with the possibility of adjusting the pH, assisting in the deposit of fine-grained tailings, and precipitating heavy metals.

1.5.2.3.1 Calcite and Magnesite Applications

Calcite is a carbonation product produced by a mineral carbonation process that uses inorganic wastes and natural rock sources such as wollastonite. The construction industry is mineral carbonation's key consumer of calcium carbonate. Calcite is also used as ground calcium carbonate (GCC) and precipitated calcium carbonate (PCC) in a range of industrial processes. PCC is pulverized limestone that ranges in particle size from a few millimeters to several microns. Also, the most important no-value use for the carbonates from mineral carbonation would be in mine reclamation projects, because of the massive amount of carbonates (Gt of magnesium carbonate) that would be produced if the mineral carbonation technology was effectively

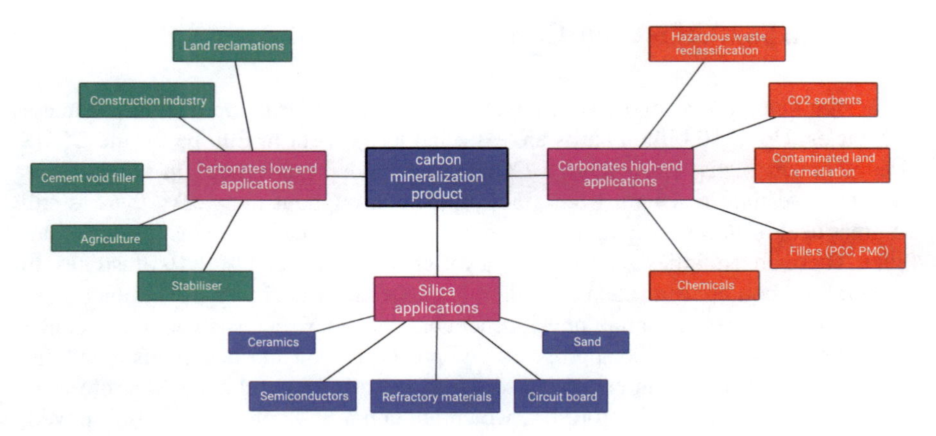

Fig. 1.18 Summary of the possible carbon mineralization product applications. (Modified after [77])

implemented. When it comes to magnesite, approximately 98% of it is converted to magnesia for conventional uses such as refractories. As a result, magnesite currently has a limited number of non-CO_2 emitting applications, such as precipitated magnesium carbonate or agricultural applications. Magnesite can be used as a building material; however, there is currently no market for it. Since the market is small, mineral carbonation magnesite is likely to be reused in large quantities for low-value applications such as land restoration programs [77].

1.5.2.3.2 Silica Applications

Mineral carbonation can produce silica byproducts as an amorphous phase, which might be utilized in the construction industry as a pozzolanic cement substitute material or as a filler. More than half of the electronic silicon raw materials marketed globally are produced in Norway. This demonstrates that mineral carbonation feedstocks are theoretically suitable for the production of high-purity silica and existing processing technologies may be used to post-process the mineralization by-products. As the electrical properties of these materials are so sensitive to impurities, it is improbable that these products can achieve such a level of purity without further post-processing [77, 106].

The applications of mineral carbonation products that do not contain calcination can be divided into three categories: low-end high-volume, high-end low-volume, and silica. Figure 1.18 shows the summary of the possible carbon mineralization product applications by this type of classification [77].

1.6 Carbon Utilization Cost

The significant costs involved in developing CCUS projects are among the major obstacles. Up to 50 billion euros are expected to be spent in Europe for the CCUS development plans. Even though CCUS projects have advanced in recent years thanks to testing in several R&D pilot projects throughout the world, there is still a great deal of uncertainty around the costs of different carbon reduction technologies. These uncertainties are caused by a variety of factors, as Fig. 1.19 illustrates. In general, carbon concentration is a significant factor in deciding the ultimate cost. Meanwhile, due to favorable physicochemical characteristics such as fine granulation, high impact level, and appropriate compositions, iron and steel industry tailings have the lowest cost. But compared to the expenses of natural sinks like reforestation, afforestation, and agroforestry, which are about \$50 per ton of CO_2 removed, these prices still appear excessive. These investments are not returned since the amount of CO_2 absorbed much exceeds any conceivable markets for CO_2, but they must be viewed as a cost to society in order to prevent unacceptably rapid climate change [107].

Costs of CO_2 reduction should be compared to potential revenue, side effects, and lifecycle studies. In many cases, corporations will need to pay extra expenses to reach net zero goals. In order to accomplish this, emitters will look for the most affordable way to reduce their emissions, aiming at the lowest cost of abatement possible. When evaluating various CO_2 lock-in potentials, it is critical to assess the broader value proposition that extends beyond managing CO_2 obligations, such as the co-benefits of CCU products. In comparison to other uses, mineral carbonation, for example, can help neutralize mining waste while also permanently storing CO_2.

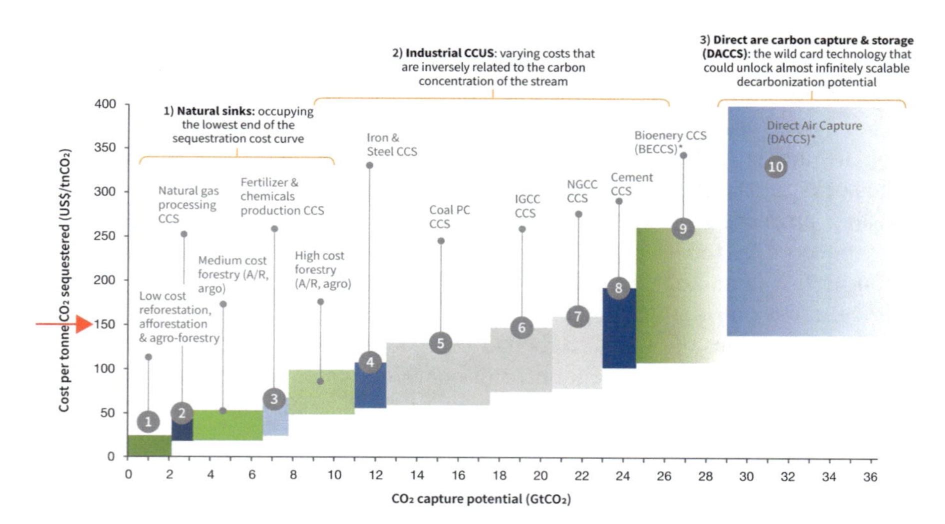

Fig. 1.19 Carbon sequestration cost curve (US\$/tn CO_2 eq) and the GHG emission abatement potential ($GtCO_2$ eq) [107]

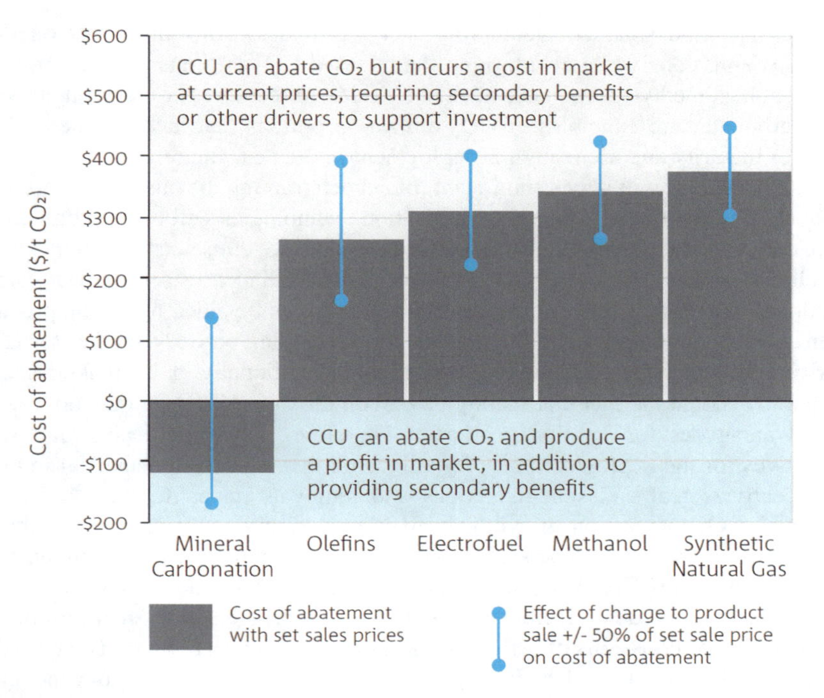

Fig. 1.20 Cost of abatement, including the effect of the product sale price on the cost of abatement [55]

Understanding the pricing variations among products might result in cost savings or the creation of new revenue sources. As shown in Fig. 1.20, the cost of abatement calculates how much each ton of CO_2 costs to avoid. The diagram examines each application of CCU best case with high partial pressure capture, where 5000 t/day of CO_2 is consumed, with the products sold at a set market price. Assumed sales prices are as follows: mineral carbonation (silica: $40/t, $MgCO_3$: $100/t), olefins ($1000/t), jet fuel ($85/bbl), methanol ($250/t), and SNG ($8/GJ). The products to the left have the lowest cost of abatement and from perspective of an emitter is likely to be pursued first [55]. As it is well demonstrated, mineral carbonation is more cost-effective than other processes.

1.7 Future Prospects

Technologies that utilize carbon have a bright future in many sectors, including the use of captured CO_2 as a renewable resource. Although it is anticipated that the market for CO_2 utilization would be small in the near future, there is already potential for the development of building materials, polymers, and industrial CO_2 use in greenhouses. CO_2 utilization for fuels in combination with renewable energy sources like wind, solar, and geothermal energy provides a convenient way to reduce net CO_2 emissions while also turning the released CO_2 into energy-rich products. Methanol, dimethyl carbonate, methane, calcium carbonates, microalgae, and polycarbonates are examples of CO_2 utilization options that have reached a high level of maturity and are ready for deployment in the near future.

Policy support will play a significant role in determining the possibilities for CO_2 utilization in the future. Many CO_2 utilization technologies will not be competitive with conventional processes unless their ability to reduce emissions is acknowledged in climate policy frameworks or there are financial incentives for lower-carbon products. The establishment of technical standards can be aided by public procurement, which can be a successful strategy for developing an early market for CO_2-derived products with provable climate benefits. Additionally, policymakers should take into account the fact that scaling up carbon dioxide utilization may not always be advantageous for stabilizing climate. Therefore, they should aim to provide incentives for the deployment of CO_2 utilization that is climate-beneficial in order to incentivize real emission reductions and removals on a life-cycle basis. For example, CO_2 conversion to synthetic fuels may significantly reduce greenhouse gas emissions by lowering dependency on fossil fuels; nevertheless, a complete life cycle assessment (LCA) is required to ensure that the process does not result in a net CO_2 emission. In order to facilitate consistent and transparent assessments of the net greenhouse gas emissions of carbon utilization technologies, research is required to create benchmark LCA of hydrogen and electricity inputs, waste gas generation, waste gas cleanup, waste gas transport, and other enabling technologies. Multiple environmental aspects of carbon utilization life cycles, including greenhouse gas emissions, water, and material use, would be included in these benchmark assessments [12, 17, 108, 109].

References

1. Romain Debarre, Prashant Gahlot, Céleste Grillet, and Mathieu Plaisant, "Carbon Capture Utilization and Storage, Towards Net-Zero," 2021.
2. M.J. Regufe, A. Pereira, A.F.P. Ferreira, A.M. Ribeiro, A.E. Rodrigues, Current developments of carbon capture storage and/or utilization–looking for net-zero emissions defined in the Paris agreement. Energies (Basel) **14**(9), 2406 (2021). https://doi.org/10.3390/en14092406
3. R. Chauvy, N. Meunier, D. Thomas, G. De Weireld, Selecting emerging CO2 utilization products for short- to mid-term deployment. Appl. Energy **236**, 662–680 (2019). https://doi.org/10.1016/j.apenergy.2018.11.096
4. S.M. Jarvis, S. Samsatli, Technologies and infrastructures underpinning future CO2 value chains: A comprehensive review and comparative analysis. Renew. Sustain. Energy Rev. **85**, 46–68 (2018). https://doi.org/10.1016/j.rser.2018.01.007
5. B.R. de Vasconcelos, J.M. Lavoie, Recent advances in power-to-X technology for the production of fuels and chemicals. Front. Chem. **7**, 1–24 (2019). https://doi.org/10.3389/fchem.2019.00392

6. K. Stangeland, D. Kalai, H. Li, Z. Yu, CO2 methanation: the effect of catalysts and reaction conditions. Energy Procedia. **105**, 2022–2027 (2017). https://doi.org/10.1016/j.egypro.2017.03.577
7. X. Su, J. Xu, B. Liang, H. Duan, B. Hou, Y. Huang, Catalytic carbon dioxide hydrogenation to methane: a review of recent studies. J. Energy Chem **25**(4), 553–565 (2016). https://doi.org/10.1016/j.jechem.2016.03.009
8. K. Ghaib, F.Z. Ben-Fares, Power-to-methane: a state-of-the-art review. Renew. Sustain. Energy Rev. **81**, 433–446 (2018). https://doi.org/10.1016/j.rser.2017.08.004
9. M. Götz et al., Renewable power-to-gas: a technological and economic review. Renew. Energy **85**, 1371–1390 (2016). https://doi.org/10.1016/j.renene.2015.07.066
10. A. O'Connell, A. Konti, M. Padella, M. Prussi, L. Lonza, *Advanced Alternative Fuels Technology Market Report 2018, EUR 29937 EN*. 2019. https://doi.org/10.2760/894775
11. M. Seemann, H. Thunman, Methane synthesis, in *Substitute Natural Gas from Waste: Technical Assessment and Industrial Applications of Biochemical and Thermochemical Processes*, (Elsevier, London, 2019), pp. 221–243. https://doi.org/10.1016/B978-0-12-815554-7.00009-X
12. E. National Academies of Sciences, *Gaseous Carbon Waste Streams Utilization: Status and Research Needs*, p. 240, 2019.
13. Y.L. Qiu, H.X. Zhong, T.T. Zhang, W. Bin Xu, X.F. Li, H.M. Zhang, Copper electrode fabricated via pulse electrodeposition: toward high methane selectivity and activity for CO2 electroreduction. ACS Catal. **7**(9), 6302–6310 (2017). https://doi.org/10.1021/acscatal.7b00571
14. D. Milani, R. Khalilpour, G. Zahedi, A. Abbas, A model-based analysis of CO2 utilization in methanol synthesis plant. J. CO2 Util. **10**, 12–22 (2015). https://doi.org/10.1016/j.jcou.2015.02.003
15. R. Guil-López et al., Methanol synthesis from CO2: a review of the latest developments in heterogeneous catalysis. Materials **12**(23), 3902 (2019). https://doi.org/10.3390/ma12233902
16. S.G. Jadhav, P.D. Vaidya, B.M. Bhanage, J.B. Joshi, Catalytic carbon dioxide hydrogenation to methanol: a review of recent studies. Chem. Eng. Res. Des. **92**(11), 2557–2567 (2014). https://doi.org/10.1016/j.cherd.2014.03.005
17. I.R. Boddula, *Carbon Dioxide Utilization to Sustainable Energy and Fuels*, vol. 2021, no. 9. in Advances in Science, Technology & Innovation (Springer, Cham, 2022). https://doi.org/10.1007/978-3-030-72877-9
18. A. Álvarez et al., Challenges in the greener production of formates/formic acid, methanol, and DME by heterogeneously catalyzed CO2 hydrogenation processes. Chem. Rev. **117**(14), 9804–9838 (2017). https://doi.org/10.1021/acs.chemrev.6b00816
19. T.A. Semelsberger, R.L. Borup, H.L. Greene, Dimethyl ether (DME) as an alternative fuel. J. Power Sources **156**(2), 497–511 (2006). https://doi.org/10.1016/j.jpowsour.2005.05.082
20. A.I. Osman, M. Hefny, M.I.A. Abdel Maksoud, A.M. Elgarahy, D.W. Rooney, Recent advances in carbon capture storage and utilisation technologies: a review. Environ. Chem. Lett. **19**(2), 797–849 (2021). https://doi.org/10.1007/s10311-020-01133-3
21. L. Vafajoo, S.H.A. Afshar, B. Firouzbakht, Developing a mathematical model for the complete kinetic cycle of direct synthesis of DME from Syngas through the CFD technique. Comput. Aid. Chem. Eng. **26**, 689–694 (2009). https://doi.org/10.1016/S1570-7946(09)70115-4
22. S. Poto, F. Gallucci, M.F.N. d'Angelo, Direct conversion of CO2 to dimethyl ether in a fixed bed membrane reactor: influence of membrane properties and process conditions. Fuel **302** (2021). https://doi.org/10.1016/j.fuel.2021.121080
23. A. Behr, K. Nowakowski, Catalytic hydrogenation of carbon dioxide to formic acid. Adv. Inorg. Chem. **66**, 223–258 (2014). https://doi.org/10.1016/B978-0-12-420221-4.00007-X

24. G.H. Gunasekar, K. Park, K.D. Jung, S. Yoon, Recent developments in the catalytic hydrogenation of CO2 to formic acid/formate using heterogeneous catalysts. Inorg. Chem. Front. **3**(7), 882–895 (2016). https://doi.org/10.1039/c5qi00231a
25. W.H. Wang, Y. Himeda, J.T. Muckerman, G.F. Manbeck, E. Fujita, CO2 hydrogenation to formate and methanol as an alternative to photo- and electrochemical CO2 reduction. Chem. Rev. **115**(23), 12936–12973 (2015). https://doi.org/10.1021/acs.chemrev.5b00197
26. Y. Huang, Y. Deng, A.D. Handoko, G.K.L. Goh, B.S. Yeo, Rational design of sulfur-doped copper catalysts for the selective electroreduction of carbon dioxide to formate. ChemSusChem **11**(1), 320–326 (2018). https://doi.org/10.1002/cssc.201701314
27. J.D. Medrano-García, R. Ruiz-Femenia, J.A. Caballero, Optimal carbon dioxide and hydrogen utilization in carbon monoxide production. J. CO2 Util. **34**, 215–230 (2019). https://doi.org/10.1016/j.jcou.2019.05.005
28. Y.H. Chan, S.N.F.S.A. Rahman, H.M. Lahuri, A. Khalid, Recent progress on CO-rich syngas production via CO2 gasification of various wastes: a critical review on efficiency, challenges and outlook. Environ. Pollut. **278** (2021). https://doi.org/10.1016/j.envpol.2021.116843
29. E. Schwab, A. Milanov, S.A. Schunk, A. Behrens, N. Schödel, Dry reforming and reverse water gas shift: alternatives for syngas production? Chem. Ing. Tech. **87**(4), 347–353 (2015). https://doi.org/10.1002/cite.201400111
30. K. Mondal, S. Sasmal, S. Badgandi, D.R. Chowdhury, V. Nair, Dry reforming of methane to syngas: a potential alternative process for value added chemicals—a techno-economic perspective. Environ. Sci. Pollut. Res. **23**(22), 22267–22273 (2016). https://doi.org/10.1007/s11356-016-6310-4
31. J.M. Lavoie, Review on dry reforming of methane, a potentially more environmentally-friendly approach to the increasing natural gas exploitation. Front. Chem. **2** (2014). https://doi.org/10.3389/fchem.2014.00081
32. A.M. Ranjekar, G.D. Yadav, Dry reforming of methane for syngas production: A review and assessment of catalyst development and efficacy. J. Indian Chem. Soc. **98**(1) (2021). https://doi.org/10.1016/j.jics.2021.100002
33. N.A.K. Aramouni, J.G. Touma, B.A. Tarboush, J. Zeaiter, M.N. Ahmad, Catalyst design for dry reforming of methane: analysis review. Renew. Sustain. Energy Rev. **82**, 2570–2585 (2018). https://doi.org/10.1016/j.rser.2017.09.076
34. A. Abdulrasheed, A.A. Jalil, Y. Gambo, M. Ibrahim, H.U. Hambali, M.Y. Shahul Hamid, A review on catalyst development for dry reforming of methane to syngas: recent advances. Renew. Sustain. Energy Rev. **108**, 175–193 (2019). https://doi.org/10.1016/j.rser.2019.03.054
35. Y. Wang et al., Syngas production via CO2 reforming of methane over aluminum-promoted NiO–10Al 2 O3 –ZrO2 catalyst. ACS Omega **6**(34), 22383–22394 (2021). https://doi.org/10.1021/acsomega.1c03174
36. A.M. Bahmanpour, M. Signorile, O. Kröcher, Recent progress in syngas production via catalytic CO2 hydrogenation reaction. Appl. Catal. B **295** (2021). https://doi.org/10.1016/j.apcatb.2021.120319
37. J. Qiao, Y. Liu, F. Hong, J. Zhang, A review of catalysts for the electroreduction of carbon dioxide to produce low-carbon fuels. Chem. Soc. Rev. **43**(2), 631–675 (2014). https://doi.org/10.1039/c3cs60323g
38. R. Küngas, Review—electrochemical CO2 reduction for CO production: comparison of low- and high-temperature electrolysis technologies. J. Electrochem. Soc. **167**(4), 044508 (2020). https://doi.org/10.1149/1945-7111/ab7099
39. C. Panzone, R. Philippe, A. Chappaz, P. Fongarland, A. Bengaouer, Power-to-Liquid catalytic CO2 valorization into fuels and chemicals: focus on the fischer-tropsch route. J. CO2 Util. **38**, 314–347 (2020). https://doi.org/10.1016/j.jcou.2020.02.009
40. J. Gorimbo et al., Lu Plot and Yao Plot: models to analyze product distribution of long-term gas-phase fischer-tropsch synthesis experimental data on an iron catalyst. Energy Fuels **31**(5), 5682–5690 (2017). https://doi.org/10.1021/acs.energyfuels.7b00388

41. H. Shibata, J.A. Moulijn, G. Mul, Enabling electrocatalytic Fischer-Tropsch synthesis from carbon dioxide over copper-based electrodes. Catal. Lett. **123**(3–4), 186–192 (2008). https://doi.org/10.1007/s10562-008-9488-3
42. S. Ma et al., Electroreduction of carbon dioxide to hydrocarbons using bimetallic Cu-Pd catalysts with different mixing patterns. J. Am. Chem. Soc. **139**(1), 47–50 (2017). https://doi.org/10.1021/jacs.6b10740
43. G. Maxwell, *Synthetic Nitrogen Products: A Practical Guide to the Products and Processes* (Springer, 2004)
44. E. Koohestanian, J. Sadeghi, D. Mohebbi-Kalhori, F. Shahraki, A. Samimi, A novel process for CO2 capture from the flue gases to produce urea and ammonia. Energy **144**, 279–285 (2018). https://doi.org/10.1016/j.energy.2017.12.034
45. E. Alper, O. Yuksel Orhan, CO2 utilization: Developments in conversion processes. Petroleum **3**(1), 109–126 (2017). https://doi.org/10.1016/j.petlm.2016.11.003
46. J.H. Meessen, H. Petersen, Urea, in *Ullmann's Encyclopedia of Industrial Chemistry*, (Wiley-VCH Verlag GmbH & Co. KGaA, Weinheim, 2000). https://doi.org/10.1002/14356007.a27_333
47. A. Gulzar, A. Gulzar, M.B. Ansari, F. He, S. Gai, P. Yang, Carbon dioxide utilization: A paradigm shift with CO2 economy. Chem. Eng. J. Adv. **3** (2020). https://doi.org/10.1016/j.ceja.2020.100013
48. Z.Z. Yang, L.N. He, J. Gao, A.H. Liu, B. Yu, Carbon dioxide utilization with C-N bond formation: carbon dioxide capture and subsequent conversion. Energy Environ. Sci. **5**(5), 6602–6639 (2012). https://doi.org/10.1039/c2ee02774g
49. C. Wu et al., Synthesis of urea derivatives from amines and CO2 in the absence of catalyst and solvent. Green Chem. **12**(10), 1811–1816 (2010). https://doi.org/10.1039/c0gc00059k
50. B. Grignard, S. Gennen, C. Jérôme, A.W. Kleij, C. Detrembleur, Advances in the use of CO2 as a renewable feedstock for the synthesis of polymers. Chem. Soc. Rev. **48**(16), 4466–4514 (2019). https://doi.org/10.1039/c9cs00047j
51. A. Ballamine, A. Kotni, J.-P. Llored, S. Caillol, Valuing CO 2 in the development of polymer materials. Sci. Technol. Energy Trans. **77**, 1 (2022). https://doi.org/10.2516/stet/2021001
52. Z. Zhang et al., Recent advances in carbon dioxide utilization. Renew. Sustain. Energy Rev. **125**, Article 109799 (2020). https://doi.org/10.1016/j.rser.2020.109799
53. R.H. Heyn, Organic carbonates, in *Carbon Dioxide Utilisation: Closing the Carbon Cycle: First Edition*, (Elsevier, 2015), pp. 97–113. https://doi.org/10.1016/B978-0-444-62746-9.00007-4
54. H.J. Ho, A. Iizuka, E. Shibata, Carbon capture and utilization technology without carbon dioxide purification and pressurization: a review on its necessity and available technologies. Ind. Eng. Chem. Res. **58**(21), 8941–8954 (2019). https://doi.org/10.1021/acs.iecr.9b01213
55. V. Srinivasan, et al., *CO2 Utilisation Roadmap*, 2021.
56. P. Chen et al., Review of biological and engineering aspects of algae to fuels approach. Int. J. Agric. Biol. Eng. **2**(4), 1–30 (2009)
57. Z. Li, X. Xin, B. Xiong, D. Zhao, X. Zhang, C. Bi, Engineering the Calvin–Benson–Bassham cycle and hydrogen utilization pathway of Ralstonia eutropha for improved autotrophic growth and polyhydroxybutyrate production. Microb. Cell. Fact. **19**(1), 228 (2020). https://doi.org/10.1186/s12934-020-01494-y
58. T.S. Wong, Carbon dioxide capture and utilization using biological systems: opportunities and challenges. J. Bioprocess Biotech. **04**(03) (2014). https://doi.org/10.4172/2155-9821.1000155
59. R.P. Sinha, D.-P. Häder, UV-protectants in cyanobacteria. Plant Sci. **174**(3), 278–289 (2008). https://doi.org/10.1016/j.plantsci.2007.12.004
60. J.M. Pisciotta, Y. Zou, I.v. Baskakov, Light-dependent electrogenic activity of cyanobacteria. PLoS One **5**(5), e10821 (2010). https://doi.org/10.1371/journal.pone.0010821
61. J.F. Kasting, J.L. Siefert, Life and the evolution of earth's atmosphere. Science (1979) **296**(5570), 1066–1068 (2002). https://doi.org/10.1126/science.1071184

62. I.M.P. Machado, S. Atsumi, Cyanobacterial biofuel production. J. Biotechnol. **162**(1), 50–56 (2012). https://doi.org/10.1016/j.jbiotec.2012.03.005
63. A. Pohlmann et al., Genome sequence of the bioplastic-producing ‘Knallgas’ bacterium Ralstonia eutropha H16. Nat. Biotechnol. **24**(10), 1257–1262 (2006). https://doi.org/10.1038/nbt1244
64. C.J. Brigham, K. Kurosawa, Bacterial carbon storage to value added products. J. Microb. Biochem. Technol. **s3**, 002 (2011). https://doi.org/10.4172/1948-5948.S3-002
65. F. Liew, A.M. Henstra, K. Winzer, M. Köpke, S.D. Simpson, N.P. Minton, Insights into CO_2 fixation pathway of *Clostridium autoethanogenum* by targeted mutagenesis. mBio **7**(3), e00427-16 (2016). https://doi.org/10.1128/mBio.00427-16
66. R. Saini, R. Kapoor, R. Kumar, T.O. Siddiqi, A. Kumar, CO2 utilizing microbes — a comprehensive review. Biotechnol. Adv. **29**(6), 949–960 (2011). https://doi.org/10.1016/j.biotechadv.2011.08.009
67. Applications for bioplastics, *European Bioplastic, Nova /institute*, 2021. https://www.european-bioplastics.org/market/applications-sectors/ Accessed 6 Aug 2022.
68. N. Hossain, T.M.I. Mahlia, R. Saidur, Latest development in microalgae-biofuel production with nano-additives. Biotechnol. Biofuels **12**(1), 125 (2019). https://doi.org/10.1186/s13068-019-1465-0
69. M. Melikoglu, V. Singh, S.-Y. Leu, C. Webb, C.S.K. Lin, Biochemical production of bioalcohols, in *Handbook of Biofuels Production*, (Elsevier, London, 2016), pp. 237–258. https://doi.org/10.1016/B978-0-08-100455-5.00009-6
70. *Negative Emissions Technologies and Reliable Sequestration*. Washington, D.C.: National Academies Press, 2019. https://doi.org/10.17226/25259.
71. B. Metz, O. Davidson, H. C. de Coninck, M. Loos, L.A. Meyer, *IPCC Special Report on Carbon Dioxide Capture and Storage*. Prepared by Working Group III of the Intergovernmental Panel on Climate Change, Cambridge/New York, 2005
72. P. Kelemen, S.M. Benson, H. Pilorgé, P. Psarras, J. Wilcox, An overview of the status and challenges of CO2 storage in minerals and geological formations. Front. Clim. **1** (2019). https://doi.org/10.3389/fclim.2019.00009
73. S.Ó. Snæbjörnsdóttir, B. Sigfússon, C. Marieni, D. Goldberg, S.R. Gislason, E.H. Oelkers, Carbon dioxide storage through mineral carbonation. Nat. Rev. Earth Environ. **1**(2), 90–102 (2020). https://doi.org/10.1038/s43017-019-0011-8
74. S.-Y. Pan et al., CO2 mineralization and utilization by alkaline solid wastes for potential carbon reduction. Nat. Sustain. **3**(5), 399–405 (2020). https://doi.org/10.1038/s41893-020-0486-9
75. C.D. Hills, N. Tripathi, P.J. Carey, Mineralization technology for carbon capture, utilization, and storage. Front. Energy Res. **8** (2020). https://doi.org/10.3389/fenrg.2020.00142
76. D. Sandalow, et al., *Carbon Mineralization Roadmap (ICEF Innovation Roadmap Project, November 2021)*. 2021
77. A. Sanna, M.R. Hall, M. Maroto-Valer, Post-processing pathways in carbon capture and storage by mineral carbonation (CCSM) towards the introduction of carbon neutral materials. Energy Environ. Sci. **5**(7), 7781 (2012). https://doi.org/10.1039/c2ee03455g
78. D. Zhang, J. Song, Mechanisms for geological carbon sequestration. Procedia IUTAM **10**, 319–327 (2014). https://doi.org/10.1016/j.piutam.2014.01.027
79. V. Romanov, Y. Soong, C. Carney, G.E. Rush, B. Nielsen, W. O'Connor, Mineralization of carbon dioxide: a literature review. ChemBioEng Rev. **2**(4), 231–256 (2015). https://doi.org/10.1002/cben.201500002
80. W. Seifritz, CO2 disposal by means of silicates. Nature **345**(6275), 486–486 (1990). https://doi.org/10.1038/345486b0
81. K.S. Lackner, C.H. Wendt, D.P. Butt, E.L. Joyce, D.H. Sharp, Carbon dioxide disposal in carbonate minerals. Energy **20**(11), 1153–1170 (1995). https://doi.org/10.1016/0360-5442(95)00071-N

82. I. Gunnarsson et al., The rapid and cost-effective capture and subsurface mineral storage of carbon and sulfur at the CarbFix2 site. Int. J. Greenhouse Gas Contr. **79**, 117–126 (2018). https://doi.org/10.1016/j.ijggc.2018.08.014
83. J.M. Matter et al., The CarbFix Pilot Project–Storing carbon dioxide in basalt. Energy Procedia **4**, 5579–5585 (2011). https://doi.org/10.1016/j.egypro.2011.02.546
84. E.H. Oelkers, S.R. Gislason, J. Matter, Mineral carbonation of CO2. Elements **4**(5), 333–337 (2008). https://doi.org/10.2113/gselements.4.5.333
85. S.R. Gíslason, H. Sigurdardóttir, E.S. Aradóttir, E.H. Oelkers, A brief history of CarbFix: Challenges and victories of the project's pilot phase. Energy Procedia **146**, 103–114 (2018). https://doi.org/10.1016/j.egypro.2018.07.014
86. "Carbfix 2," Sep. 2020
87. B.P. McGrail, F.A. Spane, E.C. Sullivan, D.H. Bacon, G. Hund, The Wallula basalt sequestration pilot project. Energy Procedia **4**, 5653–5660 (2011). https://doi.org/10.1016/j.egypro.2011.02.557
88. S.K. White et al., Quantification of CO2 mineralization at the Wallula Basalt Pilot Project. Environ. Sci. Technol. **54**(22), 14609–14616 (2020). https://doi.org/10.1021/acs.est.0c05142
89. A. Sanna, M. Uibu, G. Caramanna, R. Kuusik, M.M. Maroto-Valer, A review of mineral carbonation technologies to sequester CO_2. Chem. Soc. Rev. **43**(23), 8049–8080 (2014). https://doi.org/10.1039/C4CS00035H
90. S.P. Veetil, M. Hitch, Recent developments and challenges of aqueous mineral carbonation: a review. Int. J. Environ. Sci. Technol. **17**(10), 4359–4380 (2020). https://doi.org/10.1007/s13762-020-02776-z
91. E. R. Bobicki, Q. Liu, Z. Xu, and H. Zeng, "Carbon capture and storage using alkaline industrial wastes," Prog. Energy Combust. Sci., vol. 38, no. 2, pp. 302–320, Apr. 2012, doi: https://doi.org/10.1016/j.pecs.2011.11.002
92. S. Teir, *Fixation Of Carbon Dioxide by Producing Carbonates from Minerals and Steelmaking Slags*, Helsinki University of Technology Faculty of Engineering and Architecture, 2008
93. Y. Luo, D. He, Research status and future challenge for CO2 sequestration by mineral carbonation strategy using iron and steel slag. Environ. Sci. Pollut. Res. **28**(36), 49383–49409 (2021). https://doi.org/10.1007/s11356-021-15254-x
94. S. Eloneva, S. Teir, J. Salminen, C.-J. Fogelholm, R. Zevenhoven, Steel converter slag as a raw material for precipitation of pure calcium carbonate. Ind. Eng. Chem. Res. **47**(18), 7104–7111 (2008). https://doi.org/10.1021/ie8004034
95. S. Yadav, A. Mehra, A review on ex situ mineral carbonation. Environ. Sci. Pollut. Res. **28**(10), 12202–12231 (2021). https://doi.org/10.1007/s11356-020-12049-4
96. E. Baker, M. Davies, A. Fourie, G. Mudd, K. Thygesen, Chapter II Mine tailings facilities: overview and industry trends, in *Towards Zero Harm – A Compendium of Papers Prepared for the Global Tailings Review*, (2020)
97. G.M. Mudd, S.M. Jowitt, A detailed assessment of global nickel resource trends and endowments. Econ. Geol. **109**(7), 1813–1841 (2014). https://doi.org/10.2113/econgeo.109.7.1813
98. S.J. Gerdemann, W.K. O'Connor, D.C. Dahlin, L.R. Penner, H. Rush, Ex situ aqueous mineral carbonation. Environ. Sci. Technol. **41**(7), 2587–2593 (2007). https://doi.org/10.1021/es0619253
99. C.A. Myers, T. Nakagaki, K. Akutsu, Quantification of the CO2 mineralization potential of ironmaking and steelmaking slags under direct gas-solid reactions in flue gas. Int. J. Greenhouse Gas Contr. **87**, 100–111 (2019). https://doi.org/10.1016/j.ijggc.2019.05.021
100. Q. Zhao et al., Co-treatment of waste from steelmaking processes: steel slag-based carbon capture and storage by mineralization. Front. Chem. **8**, 571504 (2020). https://doi.org/10.3389/fchem.2020.571504
101. R.C. Sahu, R.K. Patel, B.C. Ray, Neutralization of red mud using CO2 sequestration cycle. J. Hazard Mater. **179**(1–3), 28–34 (2010). https://doi.org/10.1016/j.jhazmat.2010.02.052

102. A.R.K. Gollakota, V. Volli, C.-M. Shu, Progressive utilisation prospects of coal fly ash: a review. Sci. Total Environ. **672**, 951–989 (2019). https://doi.org/10.1016/j.scitotenv.2019.03.337
103. M. Ahmaruzzaman, A review on the utilization of fly ash. Prog. Energy Combust. Sci. **36**(3), 327–363 (2010). https://doi.org/10.1016/j.pecs.2009.11.003
104. A. Dwivedi, M. Jain, Fly ash – waste management and overview : a review, in *Recent Research in Science and Technology 2014*, (2014)
105. S.A. Ishak, H. Hashim, Low carbon measures for cement plant – a review. J. Clean Prod. **103**, 260–274 (2015). https://doi.org/10.1016/j.jclepro.2014.11.003
106. P. Gronchi, T. De Marco, L. Cassar, *Procedure for Preparing Silica from Calcium Silicate*, US6716408B1, 2004
107. Technology Brief, Carbon Capture, Use And Storage (CCUS)
108. IEA, *Putting CO2 to Use: Creating Value from Emissions*, 2019
109. C. Hepburn et al., The technological and economic prospects for CO2 utilization and removal. Nature **575**(7781), 87–97 (2019). https://doi.org/10.1038/s41586-019-1681-6

Chapter 2
The Potential of CO_2 Satellite Monitoring for Climate Governance

Fereshte Gholizadeh, Behrooz Ghobadipour, Faramarz Doulati Ardejani, Mahshad Rezaei, Aida Mirheydari, Soroush Maghsoudy, Reza Mahmoudi Kouhi, and Mohammad Milad Jebrailvand Moghaddam

2.1 Introduction

With the beginning of the industrial revolution, the heightened consumption of energy and fossil fuels has increased the emission of greenhouse gases [1]. Carbon dioxide (CO_2) and methane (CH_4) are among the most significant greenhouse gases contributing to global warming, so more than 80% of global warming is caused by the emission of these two gases [2].

Detailed studies indicate that the continued emission of greenhouse gases, particularly CO_2, will result in climate changes, including a global rise in the average surface temperature of the globe and significant regional climatic changes [3]. Figures 2.1 and 2.2 illustrate the pattern of rising the average temperature of the earth's surface and ocean as well as the atmospheric CO_2 levels.

The original version of the chapter has been revised. A correction to this chapter can be found at https://doi.org/10.1007/978-3-031-46590-1_9

F. Gholizadeh · F. Doulati Ardejani (✉) · M. Rezaei · A. Mirheydari · S. Maghsoudy
R. Mahmoudi Kouhi · M. M. Jebrailvand Moghaddam
School of Mining, College of Engineering, University of Tehran, Tehran, Iran

Climate Change Group, Mine Environment & Hydrogeology Research Laboratory (MEHR Lab.), University of Tehran, Tehran, Iran
e-mail: fereshtegholizade@ut.ac.ir; fdoulati@ut.ac.ir; Mahshadrezaei@ut.ac.ir; aida.mirheidari@ut.ac.ir; s.maghsoudy@ut.ac.ir; reza_mahmoudi@ut.ac.ir; milad.jebrailvand@ut.ac.ir

B. Ghobadipour
Climate Change Group, Mine Environment & Hydrogeology Research Laboratory (MEHR Lab.), University of Tehran, Tehran, Iran

School of Civil Engineering, Iran University of Science & Technology, Tehran, Iran

A. Ahmadian et al. (eds.), *Carbon Capture, Utilization, and Storage Technologies*, Green Energy and Technology, https://doi.org/10.1007/978-3-031-46590-1_2

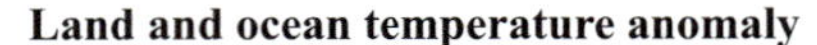

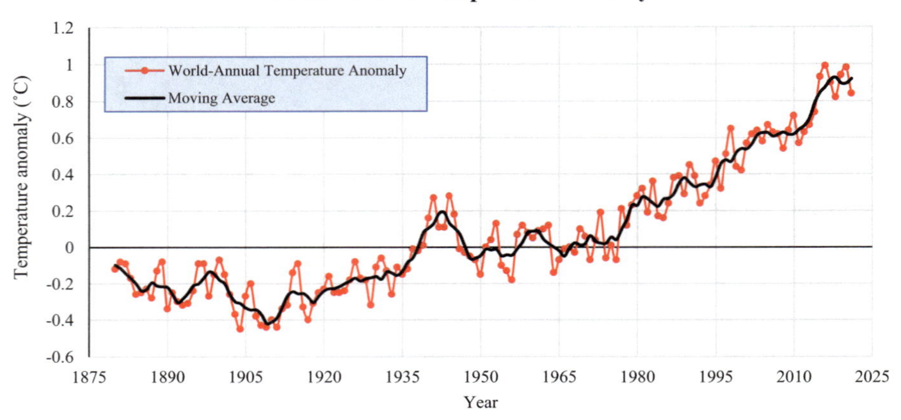

Fig. 2.1 The estimated average temperature of the earth's surface and oceans compared to the base period (1951–1980) [4]. Since 1970, the global temperature has been increasing

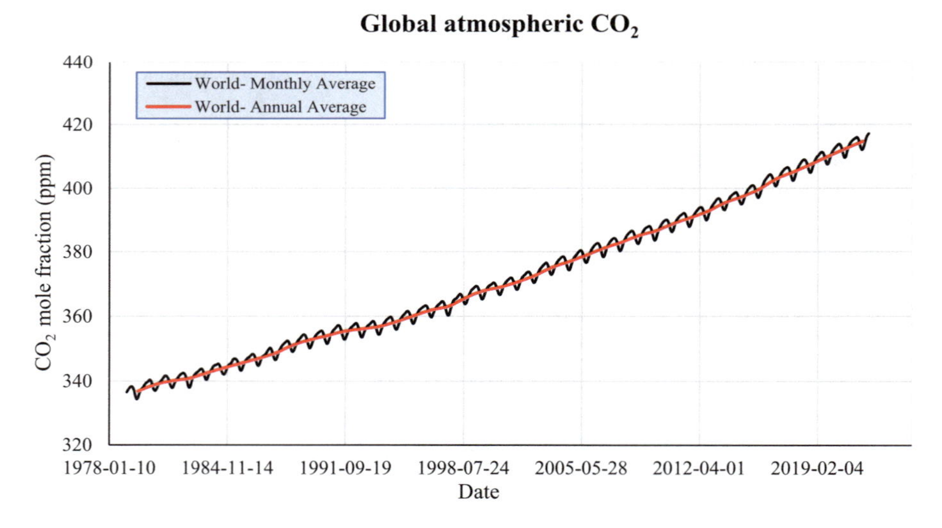

Fig. 2.2 Atmospheric CO_2 concentration (1980–2021). The graph shows the annual increase in atmospheric CO_2 concentration. Small fluctuations in monthly measurements represent seasonal changes in CO_2 emission and absorption by plants [5]

Changes in atmospheric CO_2 concentration are primarily caused by anthropogenic processes, especially the emissions of fossil fuels, photosynthesis of terrestrial ecosystems, and ocean-atmosphere and biosphere-atmosphere exchanges [6].

Large-scale anthropogenic processes cause carbon sources to be more than carbon sinks. Such activities lead to an increase in atmospheric CO_2 and affect the earth's climate system. Activities such as burning of fossil fuels, energy consumption, factories, airplanes, burning forests, decomposition of organisms, microbial respiration, animal respiration, etc., are known as sources of production in the carbon cycle. However, photosynthesis of plants, biomass, oceanic sediments, soil, etc., are sinks of this gas [7].

Since CO_2 reacts slowly with other atmospheric gases and energy sources such as solar ultraviolet radiation, most of the emitted CO_2 remains in the atmosphere for a long time (several hundred years) and thus has a long-term effect on the climate of the planet [7].

Almost half of the total CO_2 caused by anthropogenic processes is absorbed by the land and oceans, so about 25% is absorbed by vegetation on the land and 25% by the ocean, and almost 50% of CO_2 emissions remain in the atmosphere [7].

More than 70% of global anthropogenic CO_2 emissions in the atmosphere are caused by urban areas, especially megacities with more than 10 million people [8]. So the megacities of the world emit more than 20% of the global CO_2 fossil gases but only cover about 3% of the earth's surface. In addition, with the rapid growth of megacities, carbon emission has accelerated and has affected the composition of the atmosphere and climate on a global scale [9]. In the implementation of CO_2 reduction policies and goals, not only the atmospheric CO_2 concentration but also the monitoring of the CO_2 emissions is very important. Many developing countries perform poorly in providing regular and accurate CO_2 emissions reports [10]. Furthermore, CO_2 emission reports are subject to potential manipulation by various stakeholders, including CO_2 emitters and national and local governments, who may be pressured to comply with domestic CO_2 reduction policies or international obligations [11]. Satellite remote sensing by providing space observations can be widely used in environmental monitoring and CO_2 emission management [12–14].

Since 2002, satellite measurements have provided column-averaged dry-air mole fractions of carbon dioxide values (XCO_2). These data are used to investigate long-term variations in carbon emissions at both global and regional levels. In recent years, with the huge progress in the development of very accurate sensors and data gathering tools, the measurement of greenhouse gases based on satellites has facilitated the monitoring of atmospheric components [15–19]. In order to better understand how natural ecosystems and anthropogenic processes contribute to the increase in atmospheric CO_2 concentration at regional and global scales, remote sensing data has become a crucial data source [20]. Compared to ground observations, the CO_2 concentration retrieved by satellite has worldwide coverage and continuous observation features, which can reveal the spatio-temporal fluctuations of atmospheric CO_2 concentration more effectively [21].

The first satellites for monitoring greenhouse gases were launched in 2002. The first sensors and satellites that can measure CO_2 include the Atmospheric Infrared Sounder (AIRS) carried by the AQUA satellite and the Scanning Imaging Absorption Spectrometer for Atmospheric Chartography (SCIAMACHY) installed on the ENVISAT satellite. These satellites were launched in 2002 by the National Aeronautics and Space Administration (NASA) and the European Space Agency (ESA), respectively [22].

These two sensors were able to measure greenhouse gases such as CO, CO_2, CH_4 and SO_2. The SCIAMACHY instrument was shut down in April 2012 [23]; however, the AIRS instrument is still in use. AURA satellite equipped with

Tropospheric Emission Spectrometer (TES) was launched by NASA in July 2004 [24]. This satellite measures the level of greenhouse gases such as CO, CO_2, CH_4, and NH_3 in the middle troposphere. Moreover, this satellite can provide better spatial resolution than AIRS and SCIAMACHY for CO_2 monitoring (Table 2.1). ESA in cooperation with other European agencies launched METOP-A in October 2006. This satellite also measures CO, CO_2, CH_4, SO_2 and NH_3 gases, but it has poor spatial resolution ($50 \times 50\ km^2$) [25]. The greenhouse gas–observing satellite (GOSAT) was launched by Japan in January 2009 to observe $CO_{2,}$ CH4, and other gases, including H_2O and O_3 [18, 26]. The restrictions of satellite monitoring of CO_2 emissions have significantly changed since the deployment of GOSAT and its successors.

The satellite is equipped with a thermal and near-infrared sensor for carbon observation – Fourier transform spectrometer (TANSO) and a cloud and aerosol imager (CAI) [27]. GOSAT is capable of measuring CO2 with a spatial resolution of 10×10 km and a repetition period of three days [18].

Japan launched GOSAT-2 in October 2018, which is a new version of GOSAT with higher precision and cloud-free readings than the original two-meter version of GOSAT [17] owing to upgrades and enhancements made to the first two-meter version of GOSAT. NASA launched the next generation of OCO mission named OCO-3 in May 2019. The OCO-3 spectrometer has the same spectrometer as the OCO-2, which results in the same accuracy and measuring distances as the OCO-2 [15] (Table 2.1). An important feature of OCO-3 is to provide more condensed observations at sampling sites, particularly for high latitude areas, in a single day [15].

Space-based remote sensing observations with increasing spatial resolution, precision, and accuracy may be used to monitor carbon emission sources worldwide at plant [30] and city [31] scales.

GOSAT, OCO-2, GOSAT-2, TANSAT, and OCO-3 satellites with appropriate imaging features (temporal and spatial resolution) and high accuracy of the results are suitable for monitoring anthropogenic point sources of CO_2 such as cities and industrial centers worldwide [10, 19, 29].

Recent studies have used CO_2 data collected by OCO-2 to investigate anthropogenic emissions at the urban scale [32, 33]. However, OCO-2 has a long repeat cycle $\simeq$ 16 days and a relatively narrow swath band $\leq$ 10.3 km. This limitation makes it possible to identify only a small part of emissions in a city during one pass [30]. The snapshot area map (SAM) observation mode and target mode on OCO-3, which were specifically designed to measure anthropogenic emissions, have the ability to measure emissions over vast contiguous areas (approximately 80×80 km) in a pass [15, 31]. Figure 2.3 shows the time history of the beginning of activity for a number of greenhouse gas monitoring satellites.

In this study, to investigate the application of remote sensing technology in monitoring anthropogenic CO_2 emissions in urban areas, major satellites monitoring carbon emissions have been evaluated. This evaluation is based on the parameters of species observed, CO_2 sensitivity, operational period, repeat cycle, spatial

Table 2.1 The supreme CO_2 monitoring satellites (including active and inactive satellites) along with their specifications

Satellite	AQUA	ENVISAT	AURA	METOP-A	TANSAT	GOSAT	GOSAT-2	OCO-2	ISS-COC3
Sponsoring agency	NASA	ESA	NASA	ESA EUMETSAT	CAS/MOS CMA	MOE JAXA NIES	MOE JAXA NIES	NASA	NASA
Instrument	AIRS	SCIAMACHY	TES	IASI	CarbonSpec	TANSO-FTS/CAI	TANSO-FTS-2 CAI-2	Grating Spectrometer	Grating Spectrometer
Species observed	SO_2, CO, CH_4, CO_2	SO_2‘ CO‘ CH_4‘ CO2	NH_3‘ CO‘ CH_4‘ CO2	SO_2‘ CO‘ CH_4‘ CO2‘ NH_3	CO_2‘ O_2	CH_4‘ CO_2‘ O_3‘ H_2O	CH_4‘ CO_2‘ O_3‘ H_2O‘ CO‘NO_2	CO_2	CO_2
CO_2 sensitivity	Mid-troposphere	Total column	Mid-troposphere	Mid-troposphere	Total column	Total column	Total column	Total column	Total column
Operational period	2002–	2002–2012	2004–	2006–	2016–	2009–	2018–	2014–	2019–
Repeat cycle (day)	1	6	2	twice a day	16	3	6	16	[a]
Spatial resolution (km)	50 × 50	30 × 60	5.3 × 8.5	50 × 50	~ 2 × 2	~ 10 × 10	~ 10 × 10	1.29 × 2.25	1.6 × 2.2
Spectral range (μm)	3.7–15.4	0.24–2.4	3.2–15.4	3.6–15.5	0.76–2.08	0.76–14.3	0.76–14.3	0.76–2.08	0.76–2.08
CO_2 uncertainty (ppm)	1.5	14	–	2	4	4	0.5	<1	<1
References	[34]	[35]	[24]	[25]	[19]	[36]	[36]	[28]	[15]

[a]OCO-3 pointing mirror assembly system allows a much more dynamic observation-mode schedule. Therefore, the global coverage period of it is unknown [15]

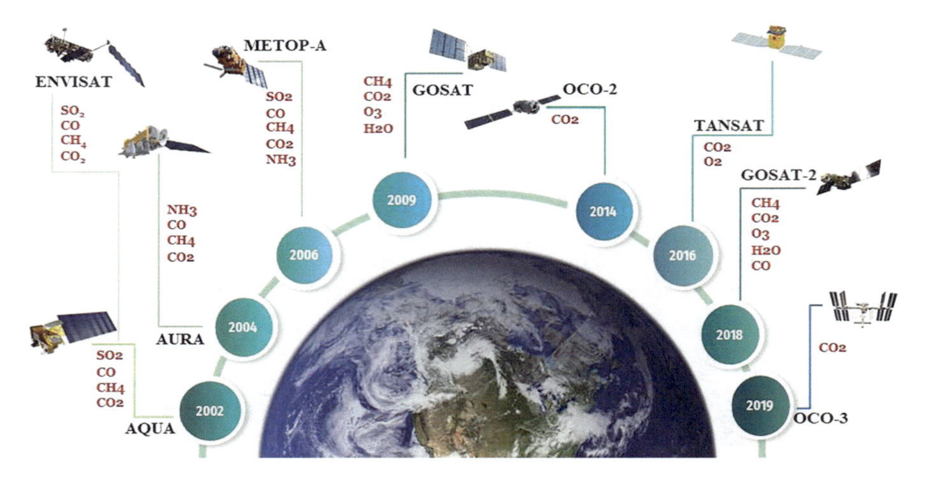

Fig. 2.3 Timeline of the launch of CO2 monitoring satellites along with the elements measured by each one

resolution, spectral range, CO_2 accuracy, observation mode, observation coverage, and observation density and gap. As a result of this evaluation, GOSAT, GOSAT-2, OCO-2, and OCO-3 satellites have been proposed and investigated more closely for detecting urban CO_2 emissions.

2.2 Satellite-Based Measurement of CO_2 Emissions

The increase of CO_2 in the atmosphere caused by anthropogenic processes has had an effect on the climate system of the earth. Therefore, measuring the concentration of CO_2 in the atmosphere along with monitoring the quantity of CO_2 emissions can be effective in managing carbon emissions.

In general, the amount of greenhouse gas emissions can be monitored and measured in two ways. The measurement, reporting, and verification (MRV) approach estimates CO_2 emissions on a regional, national, urban, or individual power plant scale. Satellite remote sensing (SRS) combined with direct monitoring may be used to monitor CO_2 emissions on a city or power plant scale.

Passive satellites record the energy resulting from the reflection of the sun rays to the earth in certain wavelengths. Carbon dioxide (CO_2) and oxygen (O_2) molecules in the atmosphere absorb the rays reflected from the ground in certain wavelengths. Therefore, the reflection received by satellite sensors has a lower amount of energy at these certain wavelengths. Through the amount of decreased energy, it is feasible to determine the amount of energy absorbed by CO_2 and O_2 molecules and their average atmospheric column [37].

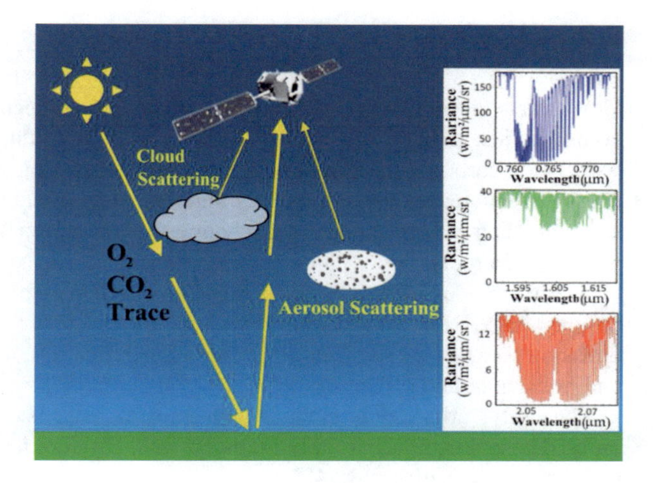

Fig. 2.4 Path of radiation and reflection of sunlight from Earth to satellite is used to calculate XCO_2

XCO_2 is defined as the column-averaged of carbon dioxide in the atmosphere. The CO_2 column abundance can be retrieved through the sunlight reflectance in the short-wavelength infrared (SWIR) band spectra. Similarly, the dry air column concentration can be estimated through the spectrum of near infrared band (NIR), which are O_2 absorbers [37]. Considering that the mole fraction of oxygen in dry air is basically constant and equal to 0.20935, XCO_2 can be obtained according to Eq. 2.1 [37].

$$\mathrm{XCO2} = 0.20935 \times \frac{\int \mathrm{NCO2\ ds}}{\int \mathrm{NO2\ ds}} \tag{2.1}$$

where

NCO_2 is the number density for CO_2 molecules, NO_2 denotes the number density for oxygen molecules, and S represents the optical path length.

The integrals are taken over the atmospheric "column" extending from the sun to the surface (Fig. 2.4) [37].

Satellite instruments estimating XCO_2 measure light reflected from the earth at specific wavelengths (mainly in the absorption spectral region of 1.6–2 μm) from the SWIR band. O_2 measurement in the absorption band of 0.76 μm is also used to estimate the optical path length [38].

2.3 City Scale CO_2 Emission Monitoring

Cities have a significant impact on climate governance. Although urban areas comprise just 2% of the earth's surface, they account for more than 70% of worldwide carbon emissions [39]. This ratio is growing as both urbanization and energy use increase [39]. For a better understanding of carbon emissions on a global scale and their influence on climate change, the evaluation of urban carbon emissions is therefore crucial.

Many previous studies for the evaluation of urban carbon emissions have focused on data based on carbon emission inventories. These data are collected from emission sectors such as electricity generation, industries, transportation, agriculture, and household and office use [3].

Currently, urban CO_2 emission inventories have been gathered in developed cities. These reports are usually prepared with different methods, so their analytical comparison is challenging [40].

Although estimating carbon emissions based on emission inventories statistically enhances our knowledge of urban carbon emissions, due to inaccuracies in the data or the data analysis method [41], it sometimes underestimates the real emissions, particularly at the urban scale [42]. Several studies have employed ground-based CO_2 measurements to decrease uncertainty in emission inventory–based approaches [43]. The spatial and geographical constraint of terrestrial CO_2 observations is one of the most significant issues with these measurements. In other words, in certain regions of the globe, such as the Middle East, there are no CO_2 measuring stations on the ground. In addition, the high costs of station building and ground measurements limit their widespread usage, particularly in less developed regions [22].

Since data from inventory analysis of carbon emission may overestimate or underestimate actual carbon emissions [44], the IPCC guidelines recommend that studies using carbon emission inventory data should validate and confirm their results using ground-based CO_2 concentration data [45].

In order to implement measures to reduce CO_2 emissions and, subsequently, to reduce climate change on the metropolitan level, there is a need for more accurate and quantitative information on CO_2 emissions. Satellite data can provide an independent, easy, and accessible method to determine the amount of the urban-level CO_2 emissions.

Many studies show the compatibility of XCO_2 satellite observation results with carbon emission inventory estimates and modeling [6, 46, 47]. As an example, the ability to measure CO_2 emissions using the SCIAMACHY meter resulted in the identification of XCO_2 values in urban areas as anthropogenic CO_2 sources [48]. Also, the use of GOSAT satellite observations to monitor urban CO_2 emissions compared to the terrestrial monitoring network (TCCON) has been able to detect changes in CO_2 emissions at a 95% confidence level [49]. OCO-2 satellite has also been able to detect XCO_2 values and enhanced anthropogenic XCO_2 in urban, suburban, and rural areas with high accuracy [6, 50, 51]. Since 2019, the OCO-3 satellite with SAM observation mode has shown a high ability to monitor CO_2

emissions, especially in metropolises. These results show that the observation of CO_2 concentration by satellites, together with ground measurements, provides reliable data sources for estimating CO_2 emissions caused by human activities and also confirming carbon emission inventories [21].

Enhanced CO_2 concentration in large cities such as Beijing, Tianjin, and Hebei regions in China, Los Angeles metropolitan area in America, Seoul metropolitan area in South Korea, and Mumbai in India has been estimated to be more than 3 ppm based on satellite remote sensing observations [48, 50, 52–55]. Overall, compared to the global annual average of XCO_2, which is estimated at 414.71 ppm in 2021 (Fig. 2.1), the increase in CO_2 concentration in cities is between 0 and 6 ppm and generally less than 3 ppm [31], which is a small contribution in measuring the total column of CO_2 concentration by satellite meters. Therefore, the accuracy of CO_2 emission estimates with the help of satellite observations in cities is highly sensitive to environmental conditions such as wind and atmospheric CO_2 transport, which should be considered before using these data to monitor urban CO_2 emission.

In general, increased CO2 levels in the point source emission sites such as urban areas, power plants, and volcanoes can be detected by removing the background concentration effect [51]. That is, the increase in CO_2 concentration can be obtained by determining the difference between the observed values at the emission source and the values in a clean area (as background) near the source. These values indicate the real contribution of cities in increasing the concentration of CO_2 in the atmosphere and are compatible with the direct emission of carbon from urban areas [31, 51].

2.4 Different Satellites for CO_2 Monitoring

In recent years, many satellites and sensors have been placed in orbit to measure greenhouse gases in the atmosphere (especially CO_2) (Fig. 2.5). Their recorded observations have a high variety in terms of spatial and temporal resolution, accuracy, collection history, etc. Comparing the spatial resolution of the satellites discussed in Sect. 2.3 shows that Aura, GOSAT, OCO-2, TANSAT, GOSAT-2, and OCO-3 satellites perform better in estimating CO_2 emissions at urban scales due to their lower spatial resolution.

To detect the concentration of CO_2 on regional and urban scales, the concentration in the lower layers of the atmosphere and areas close to the earth's surface is very important. The Aura satellite measures the concentration of CO_2 in the middle troposphere [24], but other satellites estimate the concentration of the CO_2 column in the near earth's surface [15, 19, 26, 56]. Therefore, the Aura satellite is not used to investigate anthropogenic CO_2 emissions in cities.

Both GOSAT and GOSAT-2 satellites have a spatial resolution of 10 $\times$ 10 km. GOSAT has an accuracy of 4 ppm [26]. With the improvements and progresses made on GOSAT-2, its precision has been greatly improved compared to GOSAT.

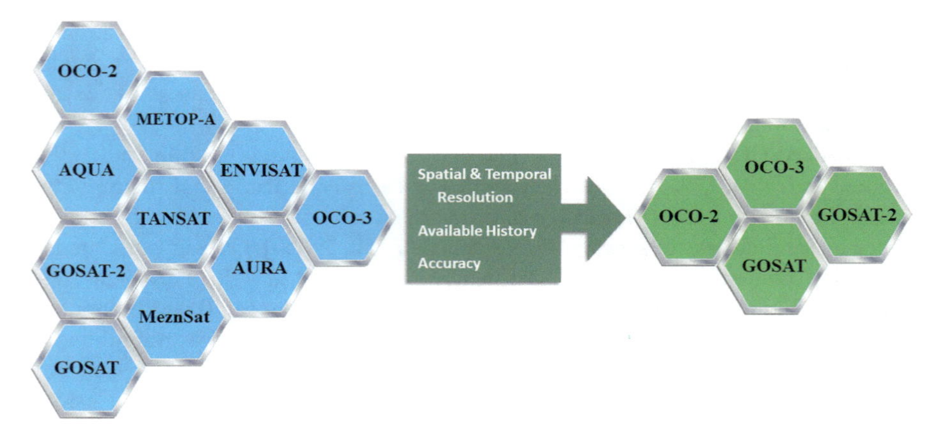

Fig. 2.5 Choosing the right satellite for urban scale anthropogenic emissions monitoring. GOSAT, OCO-2, GOSAT-2, and OCO-3 are recommended for studying urban emissions due to their time history, appropriate spatial and temporal resolution, and high accuracy

Although GOSAT has lower precision, it has a longer time history and its measurements are available since 2009 [17, 26].

OCO-2 and OCO-3 satellites have better spatial resolution than other satellites. These two satellites have recorded atmospheric CO_2 concentration since 2014 and 2019, respectively, and are currently considered among the best satellites for measuring CO_2 emissions on small scales (regional, urban, and local). The accuracy of both satellites is less than 1 ppm [15, 56].

According to the comparisons, it can be stated that OCO-2 and OCO-3 are useful in detecting small-scale CO_2 emissions due to their proper spatial resolution and high accuracy. On the other hand, GOSAT and GOSAT-2 satellites have a shorter acquisition time cycle than OCO-2 (3 and 6 days, respectively) [17, 26].

More accurate satellites such as GOSAT-2 and satellites with higher spatial resolution such as OCO-2 and OCO-3 have been launched in the last decade and do not have long-term data recording history; however, GOSAT gives older data than these satellites (since 2009) to study the history of CO_2 concentration.

Observation and collection of GOSAT data is a grid of widely spaced points with limited observations in certain areas [57]. As a result, in every time passing through the emission areas such as big cities, due to the scattering of the collection points, it covers only small parts of them.

GOSAT-2 is more focused on observing point sources in the target mode than its predecessor (GOSAT); however, due to the scattering of collected data, it cannot entirely cover urban emission zones. Compared to GOSATs, the number of observations of the total CO_2 column concentration by the OCO-2 satellite has grown dramatically, and this spacecraft gathers around one million data every day [37]. This satellite is equipped with high-precision spectrometers intended to detect smaller scales CO_2 variations. The harvesting strip on OCO-2 is about 10 kilometers wide [37]. Consequently, compared to GOSAT and GOSAT-2 satellites, it can cover a greater urban region. The OCO-3 satellite can measure more data from significant emission locations (such as cities, power plants, and volcanoes) due to its greater spatial coverage and shorter time cycle [31]. OCO-3 incorporates a method of observation intended to assess human-caused emissions. This mode, referred to as "Instantaneous Imager Area Map (SAM)," produces dense data from XCO_2 covering an area of roughly 80×80 km^2. This method of data gathering covers the majority of urban emission locations. The number of OCO-3 observations in the SAM mode, which is multi-band, is about three times more than the number of OCO-2 observations (observation in a single band) [31].

- Therefore, according to the strengths and weaknesses of each of these satellites, one can conclude that the simultaneous use of the data of these four satellites has several major advantages, which include obtaining historical data since 2009 using the GOSAT satellite.
- Study of shorter time cycles by GOSAT, GOSAT-2, and OCO-3 satellites.
- Delivering higher accuracy by using OCO-2, OCO-3, and GOSAT-2.
- Smaller spatial resolution by OCO-2, OCO-3.
- Dense spatial coverage of urban emission sources by OCO-3.

Therefore, the simultaneous use of these four satellites is suggested in order to acquire an accurate estimate for emissions with more detailed spatial resolution and better precision (Fig. 2.5).

2.4.1 GOSAT

The GOSAT satellite was launched on January 23, 2009 in cooperation with the Japan Aerospace Exploration Agency (JAXA), the Ministry of Environment (MOE), and the National Institute of Environmental Research (NIES) to study the transport mechanisms of greenhouse gases such as CO_2 and CH4 [26]. In this project, JAXA is responsible for developing the satellite and its instruments, launching the satellite and operating it (including information acquisition). NIES is responsible for data extraction and analysis. MOE is responsible for project development and financing [57]. The overall goal of launching this satellite was to help manage the environment by estimating the amount of greenhouse gases, especially CO_2, and reducing these types of gases [57].

GOSAT is the first satellite to detect CO_2 and CH4 greenhouse gas concentrations using the short-wave infrared (SWIR) band. A month after the launch of this satellite in January 2009, the operation to evaluate its performance began. The data measured by GOSAT have different levels [58].

Initially, the level one results were reported after compatibility tests with XCO_2 and XCH_4 ground measurements [58] were conducted. In February 2010, JAXA conducted early validations and released the results of level two XCO_2 and XCH_4 surface data to the public. The data was then elevated to level three. The GOSAT satellite commenced its active phase in October 2011. In the summer of 2012, this satellite officially started providing level four data [58]. NIES delivers high-resolution data from the GOSAT satellite with an inaccuracy of less than 1% every three days [59].

The orbit of the GOSAT satellite is the sun-synchronous, which moves at a distance of 666 km from the earth's surface, under an inclination angle of 98 degrees, and completes each rotation in approximately 100 minutes. This satellite crosses the equator at 13:00 ± 15 hour [58]. The data of this satellite can be accessed through the database of the European Center for Medium-Term Climate Predictions (ECMWF) / European Copernicus Climate Change Service (available at: https://cds.climat.copernicus.eu/cdsapp).

Due to the fact that the presence of fine aerosols reduces the accuracy of XCO_2 estimate, GOSAT employs the spectrometer of Cloud and Aerosol Imager (TANSO-CAI) [26] to mitigate this impact and present the XCO_2 with 1% accuracy [58].

2.4.1.1 TANSO-FTS

The GOSAT satellite is equipped with the thermal and near-infrared instrument for carbon observation (TANSO), which uses a Fourier transform spectrometer (FTS). This sensor has a wide spectrum coverage from VIS to TIR (Table 2.2) [26]. Bands 1, 2, and 3 of this meter are only able to record information during the day. However, band 4 (MWIR/TIR spectral range) does not require sunlight to collect data [58]. TANSO-FTS meter includes four spectral bands that measure O_2, O_3, CO_2, and CH4 gases (Table 2.2).

2.4.1.2 TANSO-CAI

In addition to the TANSO-FTS instrument, the GOSAT satellite is equipped with the thermal and near-infrared instrument for carbon observation (TANSO), which used a recorder tool for clouds and aerosols. TANSO-CAI is equipped with an electronic imager that makes very detailed observations of high-altitude cirrus clouds and the dispersion of suspended particles (even fine dust in near-earth regions) in the ultraviolet (UV), visible and short-wave infrared (SWIR) spectral ranges [58]. This meter is used to correct the interference of clouds and the effect of fine dust on the values of XCO_2 and XCH_4 [26] (Table 2.3). This imager has continuous spatial coverage, wider field of view, and higher spatial resolution compared to FTS [26] to detect micro-dust and cloud cover (Fig. 2.6).

Table 2.2 Spectral characteristics of the TANSO-FTS meter [60]

Spectral band no	1	2	3	4
Spectral range	VIS	SWIR	SWIR	MWIR/TIR
Coverage (μm)	0.758–0.775	1.56–1.73	1.92–2.09	5.5–14.3
Target gases	O_3	CH_4، CO_2	CO_2	CH_4، CO_2، O_3
Spectral resolution (cm^{-1})	0.5	0.2	0.2	0.2

Table 2.3 Spectral characteristics of TANSO-CAI measurement bands. This meter can detect fine particles and cloud cover and increase the accuracy of XCO_2 estimation [58]

Spectral band no	1	2	3	4
Center wavelength (μm)	0.380	0.674	0.870	1.6

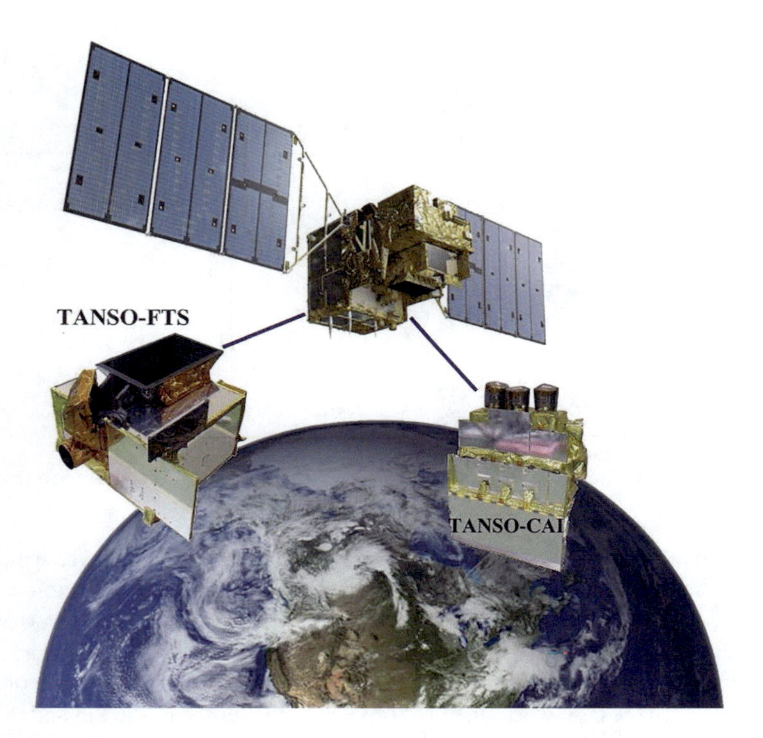

Fig. 2.6 GOSAT satellite and the position of two sensors TANSO-CAI and TANSO-FTS

2.4.1.3 GOSAT Observation Mode

The mechanism for determining the data collection points by the TANSO-FTS meter is a two-axis pointing function for cross tracks (CT) and along tracks (AT), which allows accurate observation of the ground [57]. This sensor covers a scanning angle of ±35 degrees from the nadir direction (Fig. 2.7) [57] to observe any target on Earth's surface in a three-day repetition cycle. TANSO-FTS performs CO_2 measurement in several different modes, which include

the grid point observation mode: In this mode, data collection is done as a grid with two axes along the length (AT) and width (CT) of the satellite movement. TANSO-FTS can collect observations up to an angle of ±35° in the CT direction and ±20° in the AT direction [57].

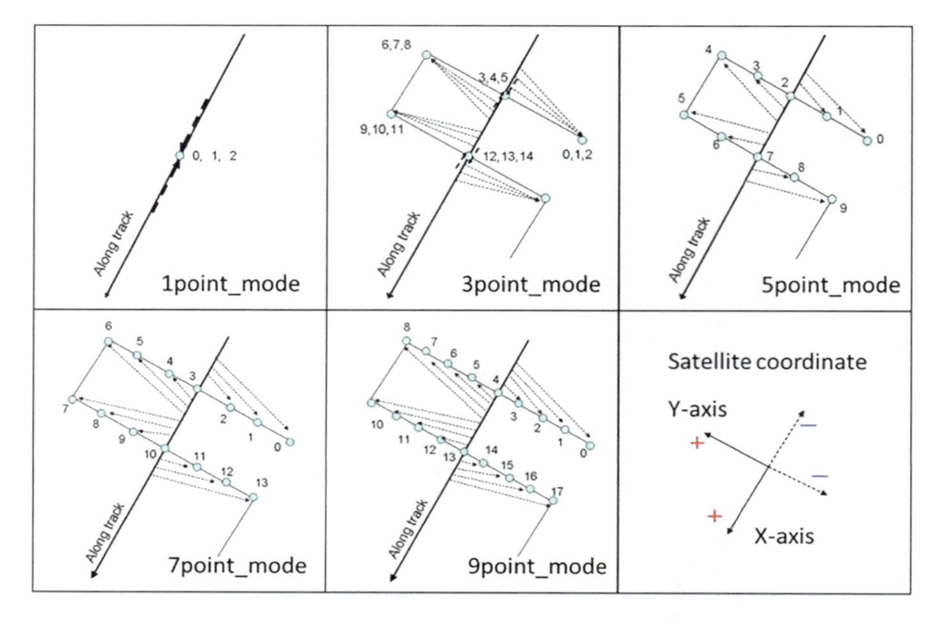

Fig. 2.7 Patterns of observation by the TANSO-FTS sensor in different fields of view [38]

The field of view of TANSO-FTS includes 1 to 9 points in the transverse direction in each sampling strip [61]. The harvesting mechanism in the CT direction can also include 1, 3, 5, 7, and 9 harvest points. The distance of these points varies from 790 (in single-point mode) to 88 km (in nine-point mode) (Fig. 2.8) [61]. The instant field of view at each sampling point is 10.5 km [57]. Figure 2.8 shows the observation geometry in the case where the field of view of the sensor is 750 km and includes 5 sampling points in the transverse direction.

TANSO-CAI also has a wide range of observations with a width of 1000 km and a viewing angle of ±35° in the CT direction [58]

Sun-glint observation mode: This type of observation usually takes place over the ocean where the surface reflection is small. In the mode of observing light reflection, the sensor observes and measures this reflection in areas where the amount of sunlight reflection is high [38].

Since the viewing angle range in the longitudinal direction (AT) is ±20°, this observation mode is limited to low and middle latitudes [38] (Fig. 2.9).

Target mode observation: TANSO-FTS meter can view the targets between ±20° in the AT direction and ±35° in the CT direction by adjusting the set of view angles. This type of observation is for validation and calibration (indirect) as well as observation of large emission sources such as large cities, active volcanoes, oil and gas fields, and landfills (Fig. 2.10) [60].

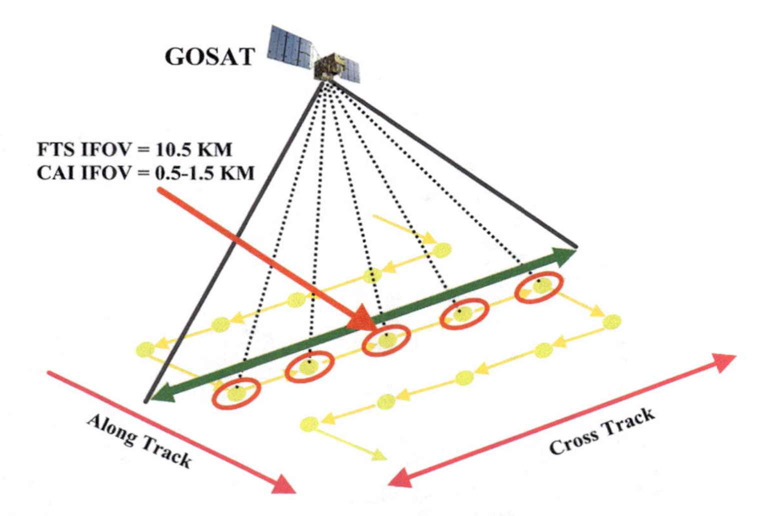

Fig. 2.8 Observation geometry by TANSO-FTS. The field of view of the sensor is 750 km and includes 5 sampling points in the transverse direction [61]

2.4.2 *GOSAT-2*

After launching the GOSAT satellite on January 23, 2009 in collaboration with the Japan Aerospace Exploration Agency (JAXA), the Ministry of the Environment (MOE), and the National Institute of Environmental Research (NIES), these three organizations launched a second satellite designated GOSAT-2. It was sent into orbit after the completion of earlier satellite operations [62]. The GOSAT satellite has successfully completed the validation and calibration procedures, resulting in enhancements such as the detection of greenhouse gases and more precise measurements of their concentration. GOSAT-2 was launched in October 2018 with upgraded FTS and CAI sensors [36] in order to enhance the precision of XCO_2 measurements. This was accomplished by enhancing the measurement accuracy and achieving an appropriate spatial resolution. This satellite was managed by the aforementioned three organizations in an effort to conduct more precise and extensive investigations and analyses using sensors with improved performance in order to monitor climate change and ultimately decrease greenhouse gas emissions [17].

A few months after the launch of this satellite, first the TANSO-CAI-2 sensor and then the TANSO-FTS-2 sensor started their activities, and as a result, the correct operation of all GOSAT-2 components was confirmed. At the beginning of February 2019, this satellite officially entered the operational stage. Level 2 data obtained from XCO_2 by the above satellite was released to the public one year after launch, in October 2019 [62].

The accuracy of observations of the GOSAT-2 satellite for measuring the concentration of CO_2, 0.5 ppm and 5 ppb for CH_4 has been reported, which has recorded more accurate observations than the GOSAT satellite (4 ppm and 34 ppb for CO_2

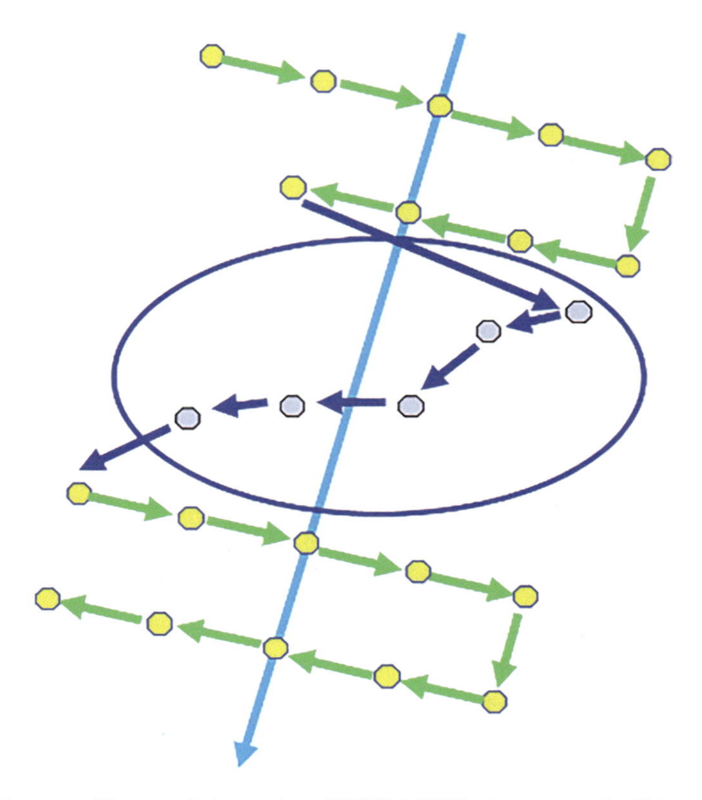

Fig. 2.9 Geometry (Geometry) observation of TANSO-FTS in the target mode. This meter adjusts the observation angles so that it observes the targets between the angles of ±20° in the AT direction and ±35° in the CT direction [38]

and CH_4, respectively) [36]. In addition, the observation period for this sun-synchronous satellite is one month, which is faster compared to the GOSAT satellite, which was three months [36]. Other successes of this satellite compared to its previous version include increasing the accuracy of estimating natural emissions, focusing more on observing target-point sources, monitoring the concentration of carbon monoxide (CO), and the use of CO and nitrogen dioxide (NO_2) to calculate human emissions [36]. The data of this satellite is available through the NIES GOSAT-2 Project database (available at: https://prdct.gosat-2.nies.go.jp).

2.4.2.1 TANSO-FTS-2

This sensor is one of the important components in the structure of the GOSAT-2 satellite, which is responsible for checking the density of greenhouse gases in the atmosphere. Compared to its previous version, TANSO-FTS-2 meter has improved features such as more accurate calibration and a faster and more intelligent pointing

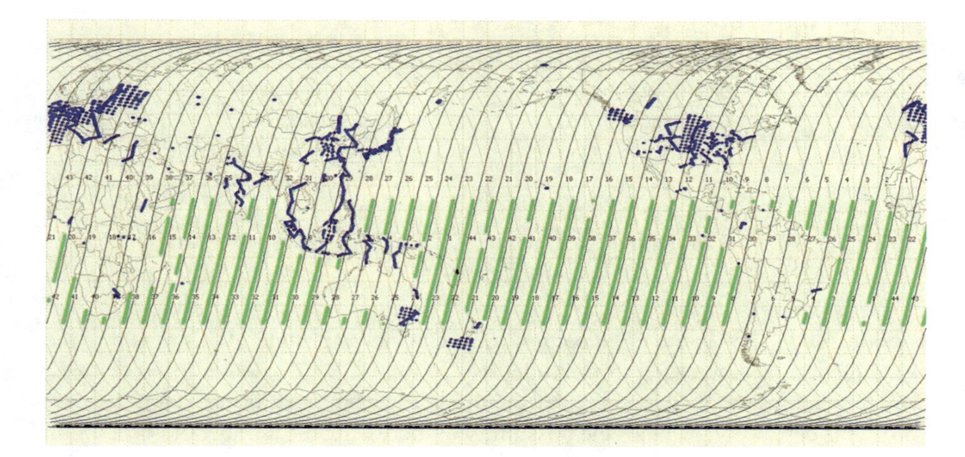

Fig. 2.10 An example of the TANSO-FTS observation pattern. The green points show the observations in the sunlight reflection mode and the blue points display the observations in the target mode [38]

Table 2.4 Comparison of specifications of GOSAT and GOSAT-2 satellites [58, 63]

Satellite mission	GOSAT	GOSAT-2
Launch date	23 January 2009	29 October 2018
Mission type	Earth observation	
Agency	MOE (Japan), JAXA, NIES (Japan)	
Measurement domain	Atmosphere, Land	
Measurement category	Aerosols, vegetation, cloud type, amount and cloud top temperature, atmospheric temperature fields, ozone, trace gases (excluding ozone)	
Gases measured	CO_2, CH_4, O_3, H_2O	CO_2, CH_4, O_3, H_2O, CO, NO_2
Instruments	TANSO-FTS, TANSO-CAI	TANSO-CAI-2, TANSO-FTS-2
Altitude (km)	666	613
Repeat cycle (day)	3	6
Spatial resolution (km)	~ 10 × 10	~ 10 × 10
Relative accuracy	4 ppm for CO_2 and 34 ppb for CH_4	0.5 ppm for CO_2 and 5 ppb for CH_4
CO_2 and CH_4 sensitivity	Total column including near surface	
Power (EOL) (KW)	3.8	5

system. Moreover, this sensor has made it possible to measure greenhouse gases in the atmosphere in cloudy conditions [17, 62]. This tool has been able to identify the sources of greenhouse gas emissions with appropriate accuracy, which has attracted the attention of many experts in the field of climate change [36]. This meter has 5 spectral bands ranging from 0.755 to 14.3 [36] (Table 2.4).

Observation modes in GOSAT-2 are largely similar to GOSAT, with the difference that TANSO-FTS-2 is able to observe targets up to an angle of ± 40° in the AT

Table 2.5 Spectral specifications of TANSO-FTS-2 meter [17, 63]

Spectral bands no	Observation bands (mμ)	Spectral resolution (cm^{-1})	Mission purpose
1	0.755–0.772	0.2	Total column O_2
2	1.563–1.695	0.2	Total column CO_2 and CH_4
3	1.923–2.381	0.2	Total column CO_2, moisture, and CO
4	5.56–8.45	0.2	CH_4 and moisture
5	8.45–14.3	0.2	Temperature profile, CO_2

direction and ± 35° in the CT direction [62]. The viewing angle of this sensor in the AT direction has increased from 20 degrees in the GOSAT satellite to 40 degrees, which increases the visible area over the ocean (measurement in the sunniest state) [63]. This capability provides the possibility of increasing the number of observation points on the ocean and contributing to global observation, including over the ocean [63].

This sensor is able to detect cloudless areas and place the scanner in the field of view (FOV). In this way, the number of cloud-free measurements without clouds increases. This capability is called intelligent pointing (IP). IP checks the position of clouds during the rotation time in the orbit and changes the field of view in the appropriate direction to obtain high-quality and cloud-free observations [62] (Table 2.5).

2.4.2.2 TANSO-CAI-2

The observation bands of this sensor capture Earth's surface with an angle of ±20° forward and backward (Fig. 2.11). This meter is used for imaging and checking the optical properties of airborne particles and clouds, as well as monitoring the state of urban and atmospheric air pollution. One of the important tasks of this meter is to separate clouds in different directions [64, 65].

Two examples of the equipment used in this satellite in the cloud separation system include:

- Using an RGB camera (on-board camera (CAM) is an RGB camera). This camera checks the presence or absence of clouds in the sampling strip of the TANSO-FTS-2 meter, and if there are clouds at the sampling point, it automatically changes the spotting angle of the meter. By using this system, the probability of obtaining cloud-free measurements is increased and, as a result, the measurement accuracy also increases [62].
- The TANSO-CAI-2 sensor has a near infrared band (0.87 μm) and a short-wave infrared band (1.60 μm), which are effective in detecting clouds [62].

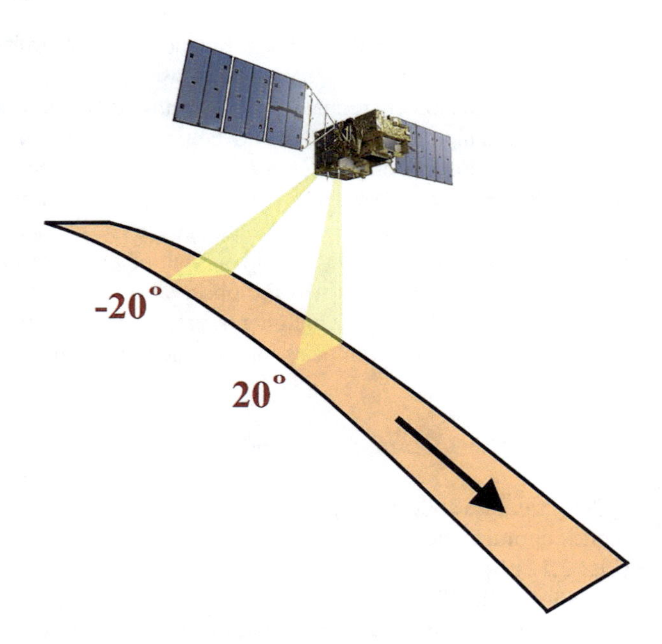

Fig. 2.11 TANSO-CAI-2 imaging range. This sensor captures Earth's surface at an angle of ±20° forward and backward [65]

2.4.3 *OCO-2*

The National Aeronautics and Space Administration (NASA) in the United States has dedicated the OCO-2 satellite to the study of carbon dioxide in the atmosphere. This satellite has the ability to provide a complete picture of human and natural carbon dioxide sources and sinks of CO_2 [37].

The OCO-2 satellite, which was launched on July 2, 2014, is actually a copy of the OCO satellite, which was decommissioned in 2009 due to technical problems. This satellite is on a sun-synchronous orbit A-Train, at an altitude of 705 km from Earth's surface, moves with an inclination angle of 2.98 degrees and crosses the equator at around 1:30 pm local time. Its rotation period is equal to 98.8 minutes and the measurement repetition period is 16 days [37].

The satellite has high-precision spectrometers designed to detect small fluctuations in XCO_2. OCO-2 measures the radiances reflected from Earth's surface in spectral bands near 0.765 μm, 1.61 μm, and 2.06 μm. OCO-2 collects data along eight parallel acquisition strips whose total width is less than or equal to 10 kilometers. OCO-2 observations have a smaller footprint from 1.29 km to 2.25 km [37].

Clouds are known to be a source of error in remote sensing measurements. The advantage of measuring with appropriate spatial resolution in OCO-2 satellite observations increases the number of cloud-free measurements and therefore allows retrieval with fewer errors [66]. This satellite has dramatically increased the number of atmospheric CO_2 observations, collecting one million measurements per day when the satellite flies over the sunlit hemisphere [37]. The data of this satellite is available through the Goddard Earth Sciences Data & Information Services Center (available at: https://disc.gsfc.nasa.gov/datasets).

This satellite collects global XCO_2 measurements with high spatial and temporal resolution and high accuracy. With the help of these measurements, regional changes of CO_2 gas in Earth's climate system can be studied more accurately. OCO-2 is the first NASA satellite specifically designed to measure atmospheric CO_2 at regional scales. This satellite can even detect changes in CO_2 concentration from smaller and isolated areas such as individual cities. One of the most interesting findings of the OCO-2 satellite in this area was the observation of a strong signal of CO_2 concentration in the Middle East, which is not present in the emission inventories, indicating that the emission inventories may be incomplete in this region [6].

2.4.3.1 OCO-2 Observation Mode

OCO-2 satellite collects data in three modes, nadir, glint, and target modes. Different modes optimize the sensitivity and accuracy of observations for specific applications (Fig. 2.12).

Nadir mode: The sensor directly observes the location of the ground under the satellite (local nadir). Rare observations provide the best spatial resolution, and in regions that are partly cloudy or topographically rugged, rare observations provide more useful XCO_2 retrievals. In addition, these observations provide more useful data at high latitudes where sunlight is low [37].

Glint mode: The meter observes a location where sunlight is directly reflected from Earth's surface (a bright spot where solar radiation is specifically reflected from the surface). The glow mode enhances the meter's ability to obtain highly accurate measurements, especially in the ocean. The signal-to-noise ratio (SNR) of glow

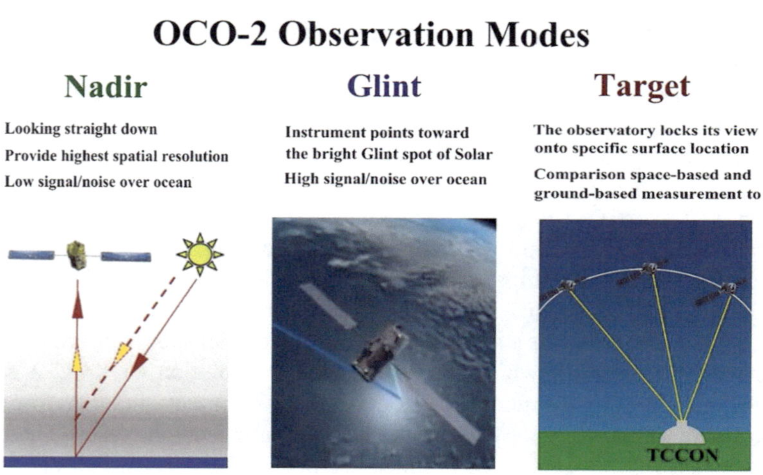

Fig. 2.12 Schematic view of OCO-2 satellite observation modes [67]

observations over dark surfaces and oceans is much higher than other regions, thus providing more accurate XCO_2 retrievals in these regions [37].

OCO-2 took its initial observations in 16-day time cycles and in only one nadir or glint mode, and after completing the cycle, the observation mode changed; so the whole land was harvested in both cases in approximately monthly time scales [37]. This approach covered oceans and continents on monthly time scales but produced 16-day long gaps in ocean coverage in a rare state. In the glow mode, this method restricted the measurements in high latitudes. In early July 2015, the observational strategy between the glint and nadir states in alternating orbits was revised. This approach now continuously covers the bright hemisphere every day [37].

Target mode: The OCO-2 meter continuously observes it when passing over a specified range (usually a XCO_2 ground measurement site). This capability provides the ability to collect thousands of measurements at locations where alternative, accurate ground-based, and airborne instruments also measure atmospheric CO_2 [37, 56].

2.4.4 OCO-3

Following the success of the OCO-2 satellite in the field of XCO_2 data acquisition, a new version called the OCO-3 satellite was sent to the International Space Station (ISS) by NASA on May 4, 2019. This satellite was sent into space with the aim of observing information about the distribution of CO_2 on Earth, which is related to the growth of urban population and changing patterns of fossil fuel combustion [68]. Although most of the components of OCO-3 are similar to OCO-2, additional equipment has been added [68].

The satellite includes three high-resolution spectrometers that collect CO_2 measurements with the accuracy, resolution, and coverage needed to assess spatial and temporal variations of CO_2 over an annual cycle to enable the understanding of sources of CO_2 production and consumption at regional scales [68]. The two spectrometers of this satellite record the wavelengths in which the absorption of reflected radiation by CO_2 molecules is strong. A third spectrometer measures the wavelengths at which the absorption of reflected radiation by oxygen molecules is strongest. By combining the data from these three spectrometers, very precise measurements of XCO_2 are obtained by this satellite [69].

The OCO-3 satellite is capable of measuring and preparing maps of XCO_2 with high resolution. Combining the observations of this satellite with the observations of the OCO-2 satellite, the most accurate maps of the effects of humans and plants on the carbon cycle and thus on the climate can be drawn [31, 51]. The data of this satellite is available through the Goddard Earth Sciences Data & Information Services Center (available at: https://disc.gsfc.nasa.gov/datasets).

Unlike other CO_2 monitoring satellites, the OCO-3 satellite can scan large continuous areas that are hotbeds of CO_2 emissions (such as cities, power plants, and volcanoes). These measurements lead to dense and microscale spatial maps of XCO2 [31].

Recent studies have shown that CO_2 data collected by the OCO-2 satellite can be used to detect urban emissions [32, 33, 70]. However, the OCO-2 satellite data have problems in providing CO_2 concentration at urban and smaller scales. OCO-2 has a long repeat cycle of 16 days. In addition, the detection band of this satellite is narrow (10.3 km). This makes only a small part of the urban emissions data to be detected during a satellite pass (if the meteorological conditions are favorable) [31]. Moreover, OCO-2 moves in a sun-synchronous orbit, which means that all observations are limited to local afternoon. The OCO-3 satellite includes an observation mode specifically designed to measure anthropogenic emissions, overcoming some of the limitations mentioned above [31].

2.4.4.1 OCO-3 Observation Mode

The OCO-3 satellite collects the solar radiation spectrum in the illuminated hemisphere in four observation modes: nadir, glow, target, and SAM [31]. Although the three modes of OCO-3's nadir, glow, and target are similar to those of OCO-2, in OCO-3, it is possible to quickly switch between different observation modes. So the transition from the rare mode to the reflection mode for this satellite takes less than a minute [69].

Snapshot area map (SAM) mode: This new feature has made it possible for the first time to detect and map the XCO_2 differences at local scales. SAM's observation mode capability can measure CO_2 emissions from small emission sources (from power plants to large urban areas) in just two minutes, while it takes several days to measure the concentration of CO_2 in such areas by OCO-2 [31, 69]. In the SAM observation mode, OCO-3 collects data in closely spaced bands. This type of acquisition leads to the production of dense and micro-scale maps of XCO_2 that cover about 80x80 square kilometers and are collected almost simultaneously, while other satellite instruments do not have this capability [31]. The number of observations of OCO-3 in the SAM mode is about three times more than the number of observations in the band-pass mode (OCO-2) [31].

During the observation of the target state, this satellite collects data in the form of sets of several swaths (usually 5 or 6 swaths), which overlap each other along the path. The overlap of these bands can cover an area of up to 20x80 km^2, which is a larger area compared to OCO-2 (20x20 km^2) [31].

Besides, OCO-3 has a larger footprint than OCO-2. The dimensions of its footprint in its nadir state are about 2.2 km in length and $\geq$ 1.6 km in width, which covers an area of 3.5 square kilometers, while the OCO-2 footprint covers an area of 3 km^2 [31].

Moreover, this satellite is able to sample megacities (one of the most important sources of greenhouse gas production and fossil fuel consumption), volcanoes, and other point emission sources and investigate the difference in CO_2 concentrations on a local scale for the first time from the space [31].

2.5 TCCON Validation

The first TCCON station was established in 2004. This station provided ground-based measurements of XCO_2 to help better understand the sources of production and consumption of CO_2 and CH_4 in Earth's atmosphere. After the construction of the first station, the TCCON network expanded in the following years. These stations provide important information on carbon-containing gases from many locations around the world [71] (Fig. 2.13).

TCCONs are a network of ground-based Fourier transform spectrometers that record the solar spectrum mainly in the near-infrared spectral region. This network can measure gases such as CO_2, CH_4, CO, N_2O, H_2O, HF, and HDO. These stations can detect small amounts of CO_2 gas in the atmospheric column at the location of the station and report the average concentration values in the entire CO_2 column [72, 73].

The main limitation of measurements at TCCON stations is that the spectrometer cannot record measurements when it is not sunny. For example, there are no measurements at night or under heavy cloud cover conditions [72]. In the absence of clouds, measurements are made at these stations approximately every two minutes. Data from each station provide information on carbon production and consumption sources at a regional scale. Additionally, by combining data from all stations, researchers can monitor carbon exchanges between the atmosphere, land, and ocean [71].

CO_2 observations from TCCON stations have shown that in the period of 2004–2014, the amount of CO_2 (XCO_2) recorded has increased by more than

Fig. 2.13 TCCON ground stations, including currently operational stations, expired stations, and stations to be launched in the future. This image was taken from the main page of the TCCON database [76]

20 ppm. This value has reached more than 400 ppm in the winter of 2014 in all stations of the Northern Hemisphere [71] and it increases on average about 2.2 ppm per year [74].

The Terrestrial Carbon Monitoring Network (TCCON) is commonly used to validate satellite XCO_2 measurements. To this end, satellites measure CO_2 in the vicinity of TCCON stations. By comparing the remotely sensed values with the values measured at TCCON stations, a complete assessment of the accuracy of CO_2 measurement by satellites is provided [75]. The GOSAT, GOSAT2, OCO-2, and OCO-3 satellites make many measurements while passing over the TCCON stations [74]. Therefore, most of the researches done by these satellites have been conducted at the metropolitan level, in cities or regions that have TCCON stations [56].

2.6 Benefits and Limitations

Remote sensing data is an effective data source for understanding the contribution of natural ecosystems and human activities to the increase of atmospheric XCO_2 at regional and global scales [20]. CO_2 monitoring satellites are different from each other in terms of specifications such as accuracy and collection history, spatial coverage, spatial and temporal resolution, etc. As mentioned, the GOSAT, GOSAT-2, OCO-2, and OCO-3 satellites are effective in detecting CO_2 emissions at regional and urban scales due to the measurement of the entire XCO_2 column, including areas close to Earth's surface [15, 18, 29, 63]. Although the acquisition history of GOSAT is longer than the other three satellites (from 2009 to present), it is less accurate (ppm 4) [36]. In terms of spatial resolution, both GOSAT and GOSAT-2 have a spatial resolution of 10 × 10 km [36], while OCO-2 and OCO-3 show a much better spatial resolution than these two satellites and therefore in detecting CO_2 emissions at urban and local scales provide more accurate measurements [29, 69].

The GOSAT and GOSAT-2 satellites have a shorter acquisition time cycle than the OCO-2 satellite (3 and 6 days, respectively) [17, 26]. OCO-3 also has a variable harvesting cycle due to being equipped with dynamic observation-mode and flexible harvesting system [15].

In terms of data spatial dispersion, the GOSAT observations are a network of scattered and limited points with large distances [57]. As a result, each time passing through the emission areas, it covers only small parts of the area in different time frames (day, week, month, and year) (Fig. 2.14).

Despite the modifications made in GOSAT-2, the number of XCO_2 observations has increased compared to the previous version, but still due to the scattering of the data collected by GOSAT-2, urban and local emission areas are not completely covered by this satellite (Fig. 2.14). The number of XCO_2 observations in the OCO-2 satellite has increased significantly compared to GOSATs [37]. Therefore, the data collected by OCO-2 is less scattered and covers much more parts of an urban area (Fig. 2.14). Like OCO-2, the OCO-3 satellite records a large number of observations. Furthermore, due to the SAM observation mode, it can identify emission areas at

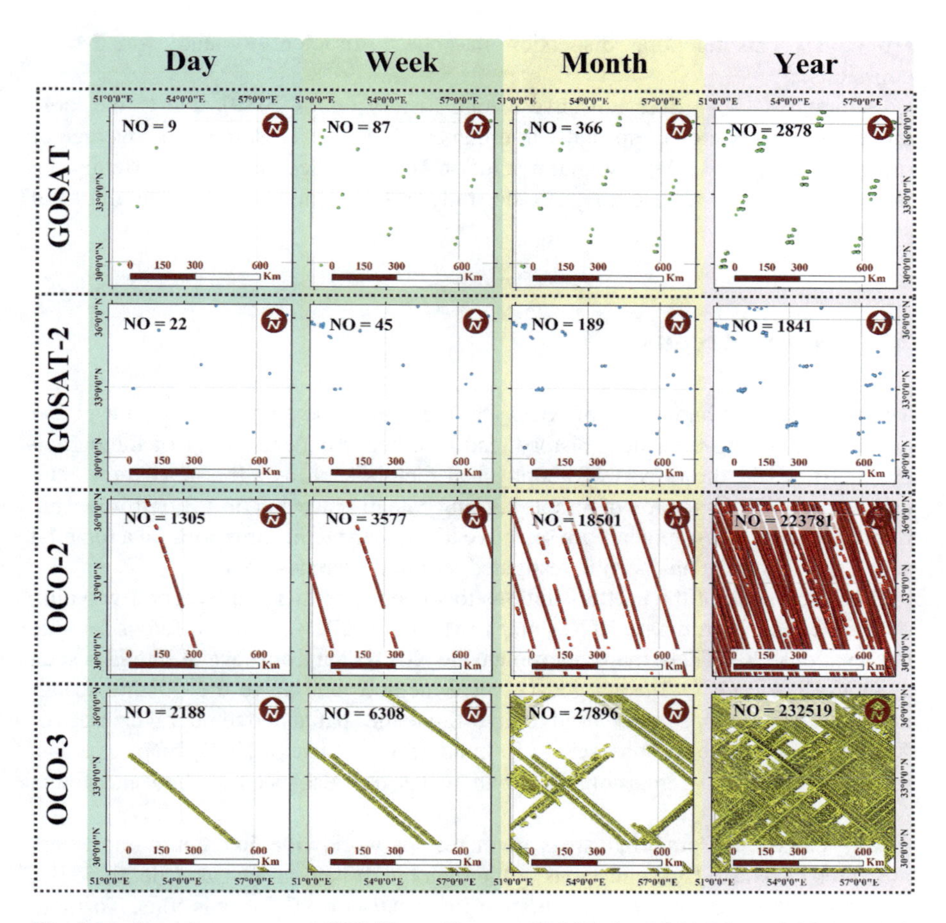

Fig. 2.14 Spatial coverage of GOSAT satellites: a: GOSAT, b: GOSAT-2, c: OCO-2, and d: OCO-3 in daily, weekly, monthly, and yearly time scales. OCO-3 and OCO-2 have the highest harvesting density. GOSAT and GOSAT-2 provide more sparse and limited measurement

small urban and local scales [31] (Fig. 2.14). The SAM observation mode is specifically designed to measure anthropogenic emissions. This capability allows OCO-3 to sample the entire surface of a large city in a short time frame and provide a large amount of XCO_2 data [31].

In general, despite the history of long-term acquisition and short-time cycle, GOSAT is less useful in small-scale emission studies due to its low accuracy, small spatial resolution, and scattering of observations.

Although GOSAT-2 has a very high accuracy, the short acquisition history (since 2019), scattering of observations, and low spatial resolution have caused limitations in the detection of regional and local emissions.

OCO-2 has advantages such as high accuracy, good spatial resolution, spatial coverage, and a relatively large number of observations and has a long-time cycle

(16 days). This temporal dispersion may be insufficient to study small-scale emissions.

Although OCO-3 does not have a long history (as of 2019), due to the large number of observations, especially in urban areas, and as a result, good coverage of urban emission areas, high spatial resolution, and high accuracy, it is currently the most suitable satellite instrument in the study of CO_2 emissions in urban and local scales.

2.7 Future Trends

The measurement and study of XCO_2 by the remote sensing method with global coverage and long-term time history lead to a better understanding of the carbon cycle. However, the effects of geophysical factors such as airborne particles and clouds cause the dispersion of remote sensing data in space and time. For this reason, remote sensing measurements are sensitive to anthropogenic emissions on a regional scale, and this issue has been investigated less in researches [20].

Therefore, one of the useful solutions to solve the spatial and temporal variation of XCO2 data is to create XCO_2 maps (Mapping-XCO_2) using different satellite observations [20]. These maps interpolate the XCO_2 data from several satellites such as GOSAT and OCO-2 using geostatistical methods and solve the existing scatter and gaps. XCO_2 maps can be used to investigate the spatial and temporal changes of XCO_2 and study the carbon cycle [77]. In addition, the use of XCO_2 maps as a data set to calculate CO_2 emissions in urban scales has also been studied and found useful [20].

CO_2 emissions from urban areas are increasing worldwide. In order to implement measures to reduce CO_2 emissions at the regional, urban, and local scale, there is a need for more accurate and quantitative information on CO_2 emissions. To more accurately estimate CO_2 increase in local and point emission areas such as urban areas, power plants, and volcanoes, the method of removing the effect of concentration values in the nearby background area is proposed [51]. So through the difference between the XCO2 values at the emission source and the values in a clean area (as background) in the vicinity of the source, it is possible to estimate the amount of CO_2 emission at the production source [31]. An area where CO_2 emissions are negligible and not affected by pollutant emissions in its vicinity is called a clean area. Due to atmospheric disturbances, in order to estimate the amount of CO_2 emissions at the production source, the satellite observations of this area and the background area must have been taken almost at the same time.

The accuracy of CO_2 emission estimation by the remote sensing method is very sensitive to environmental conditions such as wind and atmospheric transport of CO_2. These data are affected by clouds, suspended particles, and weather conditions; and for this reason, they may not be of very good quality. For this reason, it is recommended to study the environmental conditions such as topography and wind condition of the area along with CO_2 emission estimation studies. Furthermore, validation with ground-based measurements such as XCO_2 measurements at TCCON stations can be useful to achieve more reliable results at local and urban scales [51]. Unfortunately, only a few cities around the world have ground CO_2 measuring stations. Therefore, satellite data with high accuracy and high spatial and temporal resolution, such as OCO-2 data as well as OCO-3 and GOSAT-2 data, are suggested for detailed and deeper studies [51].

Currently, CO_2 measurement in the SAM mode and the target mode by OCO-3 is an innovative dataset for urban and point scale carbon studies [31]. In order to more closely study the carbon cycle and CO_2 emissions, extensive space missions are planned in the future. With dense spatial coverage and short time repetition cycles, these spacecrafts can play an important role in quantifying CO_2 emissions in urban areas, potentially monitoring the effectiveness and progress of CO_2 emission reduction policies [31]. Table 2.6 presents a number of these space missions showing their applications and characteristics.

2.8 Conclusion

The increase in the population of cities has led to an inhance of man-made CO_2 emissions and an increase in global temperature, especially in recent decades. By sending satellites that monitor greenhouse gases, monitoring the emission and measuring the CO_2 concentration have become more possible. In this research, the application of remote sensing technology to monitor CO_2 emissions in cities has been investigated. Therefore, CO_2 monitoring satellites (AQUA, ENVISAT, AURA, METOP-A, GOSAT, OCO-2, TANSAT, GOSAT-2, and OCO-3) were evaluated and compared in terms of anthropogenic CO_2 monitored in cities. This comparison is based on the parameters of species observed, CO_2 sensitivity, operational period, repeat cycle, spatial resolution, spectral range, CO_2 accuracy, observation mode, observation coverage, and observation density and gap. GOSAT, GOSAT-2, OCO-2, and OCO-3 satellites are effective in detecting CO_2 emissions at regional and urban scales due to the measurement of the entire XCO_2 column, including areas close to Earth's surface. For studies that require a longer history, GOSAT can provide valuable information. The increase in CO_2 concentration in cities is between 0 and 6 ppm and generally less than 3 ppm. Therefore, observations of GOSAT-2, OCO-2, and OCO-3, which have a sampling accuracy of less than 1 ppm, can be used to monitor the amount of anthropogenic emissions in cities. OCO-2 and OCO-3 have much better spatial resolution than other satellites. Therefore, they provide more accurate measurements in detecting CO_2 emissions at urban and local scales. OCO-2 and OCO-3 cover more cities and more parts of these urban areas due to the large number of XCO_2 observations. The SAM observation mode on OCO-3 is specifically designed to measure anthropogenic emissions. This capability allows OCO-3 to survey the entire surface of a large city in a short period of time. Although OCO-3 does not have a long history of data acquisition (since 2019), due to the large number of observations, especially in urban areas, spatial resolution, and

Table 2.6 Future planned space missions, their characteristics, and applications [78]

Mission	Tansat-2	GeoCarb	Carbon mapper (CM)	MicroCarb	GOSAT-GW (TANSO-3)	Sentinel-Co_2M-A	CO_2Image
Instruments	CAPI2, LASHIS TANSAT-2 Pollution	Scanning Spectrometer (GeoCarb)	hyperspectral imaging spectrometer	Microcarb	AMSR3-TANSO-3	CLIM, CO2I, MAO	COSIS
Agency	IAMCAS, SARI, STU	NASA	NASA, Arizona State University, JPL, ASU	CNES, UKSA	JAXA, MOE, NIES	COM-EUMETSAT	DLR
Launch date	2022	2022	2023	23-Dec	24-Mar	25-Dec	26-May
Type	Sun-synchronous	–	–	Sun-synchronous	Sun-synchronous	Sun-synchronous	Sun-synchronous
Altitude (km)	600	35400	–	650	666	735	500
Period (mins)	–	–	–		98.18	99.5	92.2
Repeat cycle (days)	–	1	–	7	3	11	–
Waveband	–	0.76, 1.61, 2.06, and 2.32 μm	–	10.5–17.6 nm	6.925–183.3 gh	405–2095 nm	~1.3 μm–~3.0 μm
Spatial resolution	–	3×6 km^2	–	2×2 km^2	5–50 km	4 km^2	–
Accuracy	–	>2.7 ppm	–	1 ppm		0.7	–
Applications	6 satellite to measure CO_2, O2, CH4, and CO emissions	4-channel slit imaging spectrometer to measure XCO2, XCO, CH4, O2, and SIF	Pinpoint, quantify, and track point-source methane and CO2 emissions.	Measure CO_2 concentration	Observation of GHG and study of water cycle mechanisms	Measure anthropogenic CO_2 emissions at country and megacity scales.	Measure CO_2 emissions from localized sources

high accuracy, it is currently the most suitable satellite instrument in the study of CO_2 emissions at urban and local scales.

The use of remote sensing satellites for emission monitoring is only able to estimate the average column of the total mole fraction of CO_2, including the atmospheric region close to Earth's surface, and cannot determine the amount of CO_2 gas emissions. However, they can monitor emission changes with high accuracy by removing the background. In recent years, many advances have been made in the field of satellite monitoring of greenhouse gas emissions, and many satellites have been placed in orbit for this purpose. In addition, in future missions, satellites such as Tansat-2, GeoCarb, Carbon mapper (CM), MicroCarb, GOSAT-GW (TANSO-3), Sentinel-Co2M-A, and CO_2 Image can also play an important role in monitoring urban emissions in line with the policy of reducing CO_2 emissions.

References

1. J. E. Hansen, *Global Warming: The Complete Briefing,* 2nd ed, vol. 30(3), 1998.
2. P. Prasad, S. Rastogi, R.P. Singh, Study of satellite retrieved CO_2 and CH_4 concentration over India. Adv. Sp. Res. **54**(9), 1933–1940 (2014). https://doi.org/10.1016/j.asr.2014.07.021
3. R.K. Pachauri, L.A. Meyer, Climate change 2014: synthesis report. in *Contributed Working Groups I, II III to Fifth Assessment Report* (Intergovernmental Panel on Climate Change, 2014).
4. N.J.L. Lenssen et al., Improvements in the GISTEMP uncertainty model. J. Geophys. Res. Atmos. **124**(12), 6307–6326 (2019)
5. NOAA, National oceanic and atmospheric administration (noaa), 2021. https://gml.noaa.gov/ccgg/trends/. Accessed 1 May 2022.
6. J. Hakkarainen, I. Ialongo, J. Tamminen, Direct space-based observations of anthropogenic CO2 emission areas from OCO-2. Geophys. Res. Lett. **43**(21), 11,400–11,406 (2016). https://doi.org/10.1002/2016GL070885
7. H. Hanson, K. Yuen, D. Crip, Orbiting carbon observatory-2: observing CO2 from space. Earth Obs. **26**, 4–11 (2014)
8. United Nations/Department of Economic and Social Affairs, *World Urbanization Prospects: The 2018 Revision*, 2019. https://doi.org/10.18356/b9e995fe-en.
9. T.F. Stocker et al., *Climate Change 2013: The physical science basis. contribution of working group I to the fifth assessment report of IPCC the intergovernmental panel on climate change* (Cambridge University Press, 2014)
10. National Research council, *Verifying greenhouse gas emissions: methods to support international climate agreements* (National Academies Press, 2010)
11. Y. Xu, Environmental Policy and Air Pollution in China. Environ. Policy Air Pollut. China (2020). https://doi.org/10.4324/9780429452154
12. V. De Sy et al., Synergies of multiple remote sensing data sources for REDD+ monitoring. Curr. Opin. Environ. Sustain. **4**(6), 696–706 (2012). https://doi.org/10.1016/j.cosust.2012.09.013
13. H.K. Gibbs, S. Brown, J.O. Niles, J.A. Foley, Monitoring and estimating tropical forest carbon stocks: making REDD a reality. Environ. Res. Lett. **2**(4), 45023 (2007)
14. W. Turner, S. Spector, N. Gardiner, M. Fladeland, E. Sterling, M. Steininger, Remote sensing for biodiversity science and conservation. Trends Ecol. Evol. **18**(6), 306–314 (2003)
15. A. Eldering, T.E. Taylor, C.W. O'Dell, R. Pavlick, The OCO-3 mission: Measurement objectives and expected performance based on 1 year of simulated data. Atmos. Meas. Tech. **12**(4), 2341–2370 (2019). https://doi.org/10.5194/amt-12-2341-2019

16. B. Connor et al., Quantification of uncertainties in OCO-2 measurements of XCO2: simulations and linear error analysis, **2009**, 5227–5238, 2016. https://doi.org/10.5194/amt-9-5227-2016.
17. R. Glumb, G. Davis, C. Lietzke, The TANSO-FTS-2 instrument for the GOSAT-2 greenhouse gas monitoring mission. Int. Geosci. Remote Sens. Symp., 1238–1240 (2014). https://doi.org/10.1109/IGARSS.2014.6946656
18. A. Kuze, H. Suto, M. Nakajima, T. Hamazaki, Thermal and near infrared sensor for carbon observation Fourier-transform spectrometer on the Greenhouse Gases Observing Satellite for greenhouse gases monitoring. Appl. Opt. **48**(35), 6716–6733 (2009). https://doi.org/10.1364/AO.48.006716
19. Y. Liu et al., The TanSat mission: preliminary global observations. Sci. Bull. **63**(18), 1200–1207 (2018). https://doi.org/10.1016/j.scib.2018.08.004
20. M. Sheng, L. Lei, Z.C. Zeng, W. Rao, S. Zhang, Detecting the responses of co2 column abundances to anthropogenic emissions from satellite observations of gosat and oco-2. Remote Sens. **13**(17) (2021). https://doi.org/10.3390/rs13173524
21. S. Yang, L. Lei, Z. Zeng, Z. He, H. Zhong, An assessment of anthropogenic CO2 emissions by satellite-based observations in China. Sensors **19**(5), 1118 (2019)
22. G. Pan, Y. Xu, J. Ma, The potential of CO_2 satellite monitoring for climate governance: a review. J. Environ. Manage. **277**, 111423, 2021. https://doi.org/10.1016/j.jenvman.2020.111423
23. D.G. Streets et al., Emissions estimation from satellite retrievals: A review of current capability. Atmos. Environ. **77**, 1011–1042 (2013). https://doi.org/10.1016/j.atmosenv.2013.05.051
24. R. Beer, T.A. Glavich, D.M. Rider, Tropospheric emission spectrometer for the earth observing system's aura satellite. Appl. Opt. **40**(15), 2356 (2001). https://doi.org/10.1364/ao.40.002356
25. C. Clerbaux et al., Monitoring of atmospheric composition using the thermal infrared IASI/MetOp sounder. Atmos. Chem. Phys. **9**(16), 6041–6054 (2009)
26. T. Yokota et al., Global concentrations of CO2 and CH4 retrieved from GOSAT: First preliminary results. Sci. Online Lett. Atmos. **5**(1), 160–163 (2009). https://doi.org/10.2151/sola.2009-041
27. T. Hamazaki, A. Kuze, K. Kondo, Sensor system for Greenhouse Gas Observing Satellite (GOSAT). Infrared Spaceborne Remote Sens XII **5543**, 275 (2004). https://doi.org/10.1117/12.560589
28. D. Crisp et al., The orbiting carbon observatory (OCO) mission. Adv. Sp. Res. **34**(4), 700–709 (2004)
29. D. Crisp et al., The on-orbit performance of the Orbiting Carbon Observatory-2 (OCO-2) instrument and its radiometrically calibrated products. Atmos. Meas. Tech. **10**(1), 59–81 (2017)
30. R. Nassar, T.G. Hill, C.A. McLinden, D. Wunch, D.B.A. Jones, D. Crisp, Quantifying CO_2 emissions from individual power plants from space. Geophys. Res. Lett. **44**(19), 10,045–10,053 (2017). https://doi.org/10.1002/2017GL074702
31. M. Kiel et al., Urban-focused satellite CO2 observations from the Orbiting Carbon Observatory-3: a first look at the Los Angeles megacity. Remote Sens. Environ. **258**(August), 2021 (2020). https://doi.org/10.1016/j.rse.2021.112314
32. D. Wu, J.C. Lin, T. Oda, E.A. Kort, Space-based quantification of per capita CO2 emissions from cities. Environ. Res. Lett. **15**(3) (2020). https://doi.org/10.1088/1748-9326/ab68eb
33. X. Ye et al., Constraining Fossil Fuel CO2 Emissions From Urban Area Using OCO-2 Observations of Total Column CO2. J. Geophys. Res. Atmos. **125**(8), 1–29 (2020). https://doi.org/10.1029/2019JD030528
34. H.H. Aumann et al., AIRS/AMSU/HSB on the Aqua mission: Design, science objectives, data products, and processing systems. IEEE Trans. Geosci. Remote Sens. **41**(2), 253–264 (2003)
35. H. Bovensmann et al., A remote sensing technique for global monitoring of power plant CO_2 emissions from space and related applications. Atmos. Meas. Tech. **3**(4), 781–811 (2010). https://doi.org/10.5194/amt-3-781-2010
36. M. Nakajima, A. Kuze, H. Suto, The current status of GOSAT and the concept of GOSAT-2. Sensors, Syst. Next-Generation Satell XVI **8533**, 21–30 (2012)

37. D. Crisp, Measuring atmospheric carbon dioxide from space with the Orbiting Carbon Observatory-2 (OCO-2). Earth Obs. Syst. **9607**, 960702 (2015)
38. EORC JAXA, TANSO-FTS – Fourier Transform Spectrometer. https://www.eorc.jaxa.jp/GOSAT/instrument_1.html. Accessed 18 Nov 2022.
39. UN-HABITAT, *Cities for All: Bridging the Urban Divide – State of the World's Cities 2010/2011 by UN-HABITAT*. 2011.
40. L. Kamal-Chaoui, A. Robert, *Competitive Cities and Climate Change*, 2009.
41. K. Rypdal, W. Winiwarter, Uncertainties in greenhouse gas emission inventories - Evaluation, comparability and implications. Environ. Sci. Policy **4**(2–3), 107–116 (2001). https://doi.org/10.1016/S1462-9011(00)00113-1
42. N. Bader, R. Bleischwitz, Measuring urban greenhouse gas emissions: the challenge of comparability. Cities Clim. Chang. **2**(3), 1–15 (2009)
43. L.R. Hutyra et al., Urbanization and the carbon cycle: current capabilities and research outlook from the natural sciences perspective. Earth's Futur. **2**(10), 473–495 (2014)
44. F.M. Bréon et al., An attempt at estimating Paris area CO_2 emissions from, 1707–1724, 2015, https://doi.org/10.5194/acp-15-1707-2015.
45. Z. Zhongming, L. Linong, Y. Xiaona, Z. Wangqiang, L. Wei, *2019 Refinement to the 2006 IPCC Guidelines for National Greenhouse Gas Inventories*, 2019.
46. R. Janardanan et al., *Comparing GOSAT Observations of Localized CO_2 Enhancements by Large Emitters with Inventory-Based Estimates*, June 2009, 3486–3493, 2016, https://doi.org/10.1002/2016GL067843.Received.
47. J. Hakkarainen, I. Ialongo, S. Maksyutov, *Analysis of Four Years of Global XCO 2 Anomalies as Seen by Orbiting Carbon Observatory-2*, 1–20, 2019, https://doi.org/10.3390/rs11070850.
48. O. Schneising, J. Heymann, M. Buchwitz, M. Reuter, H. Bovensmann, J.P. Burrows, Anthropogenic carbon dioxide source areas observed from space: assessment of regional enhancements and trends. Atmos. Chem. Phys. **13**(5), 2445–2454 (2013). https://doi.org/10.5194/acp-13-2445-2013
49. K. McKain, S.C. Wofsy, T. Nehrkorn, J. Eluszkiewicz, J.R. Ehleringer, B.B. Stephens, Assessment of ground-based atmospheric observations for verification of greenhouse gas emissions from an urban region. Proc. Natl. Acad. Sci. U. S. A. **109**(22), 8423–8428 (2012). https://doi.org/10.1073/pnas.1116645109
50. F.M. Schwandner et al., Spaceborne detection of localized carbon dioxide sources, *Science (80–.).*, **358**(6360), 2017, https://doi.org/10.1126/science.aam5782.
51. C. Park, S. Jeong, H. Park, J. Yun, J. Liu, Evaluation of the potential use of satellite-derived XCO2 in detecting CO2 enhancement in megacities with limited ground observations: a case study in seoul using orbiting carbon observatory-2. Asia-Pacific J. Atmos. Sci. **57**(2), 289–299 (2021). https://doi.org/10.1007/s13143-020-00202-5
52. A. Eldering et al., The Orbiting Carbon Observatory-2 early science investigations of regional carbon dioxide fluxes," *Science (80–).*, **358**(6360), eaam5745, 2017.
53. E.A. Kort, C. Frankenberg, C.E. Miller, T. Oda, Space-based observations of megacity carbon dioxide. Geophys. Res. Lett. **39**(17), 1–5 (2012). https://doi.org/10.1029/2012GL052738
54. G. Keppel-Aleks, P.O. Wennberg, C.W. O'Dell, D. Wunch, Towards constraints on fossil fuel emissions from total column carbon dioxide. Atmos. Chem. Phys. **13**(8), 4349–4357 (2013)
55. L. Lei et al., Assessment of atmospheric CO2 concentration enhancement from anthropogenic emissions based on satellite observations. Kexue Tongbao/Chinese Sci. Bull. **62**(25), 2941–2950 (2017). https://doi.org/10.1360/N972016-01316
56. D. Wunch et al., Comparisons of the orbiting carbon observatory-2 (OCO-2) X CO 2 measurements with TCCON. Atmos. Meas. Tech. **10**(6), 2209–2238 (2017)
57. T. Hamazaki, Y. Kaneko, A. Kuze, K. Kondo, Fourier transform spectrometer for Greenhouse Gases Observing Satellite (GOSAT). Enabling Sens. Platf. Technol. Spaceborne Remote Sens. **5659**, 73 (2005). https://doi.org/10.1117/12.581198

58. *GOSAT/IBUKI Data Users Handbook 1st Edition*, no. March. Satellite Applications and Promotion Centre, Space Applications Mission Directorate, Japan Aerospace Exploration Agency, Japan, 2011.
59. A. Kuze, T. Urabe, H. Suto, Y. Kaneko, T. Hamazaki, The instrumentation and the BBM test results of thermal and near-infrared sensor for carbon observation (TANSO) on GOSAT. Infrared Spaceborne Remote Sensing XIV **6297**, 138–145 (2006)
60. K. Shiomi et al., GOSAT level 1 processing and in-orbit calibration plan. Sensors, Syst. Next-Generation Satell. XII **7106**, 71060O (2008). https://doi.org/10.1117/12.800278
61. T. Hamazaki, *Greenhouse Gases Observation from Space: Overview of TANSO and GOSAT*, 2008. https://doi.org/10.1117/12.2308255.
62. *GOSAT-2/IBUKI-2 Data Users Handbook 1st Edition*. Satellite Applications and Promotion Centre, Space Applications Mission Directorate, Japan Aerospace Exploration Agency, Japan, 2020.
63. M. Nakajima, H. Suto, K. Yotsumoto, T. Miyakawa, K. Shiomi, GOSAT-2: Development Status of the mission instruments, in *EGU General Assembly Conference Abstracts*, 2015, p. 7731.
64. Y. Oishi, H. Ishida, T.Y. Nakajima, R. Nakamura, T. Matsunaga, The impact of different support vectors on GOSAT-2 CAI-2 L2 cloud discrimination. Remote Sens. **9**(12), 1–13 (2017). https://doi.org/10.3390/rs9121236
65. Y. Oishi, T.Y. Nakajima, T. Matsunaga, Difference between forward-and backward-looking bands of GOSAT-2 CAI-2 cloud discrimination used with Terra MISR data. Int. J. Remote Sens. **37**(5), 1115–1126 (2016)
66. C.W. O'Dell et al., Improved retrievals of carbon dioxide from Orbiting Carbon Observatory-2 with the version 8 ACOS algorithm. Atmos. Meas. Tech. **11**(12), 6539–6576 (2018). https://doi.org/10.5194/amt-11-6539-2018
67. NASA/JPL, "Orbiting Carbon Observatory-2," *NASA Jet Propulsion Laboratory*. https://ocov2.jpl.nasa.gov/. Accessed 20 Nov 2022.
68. NASA/JPL, "Orbiting Carbon Observatory-3," *NASA Jet Propulsion Laboratory*. https://ocov3.jpl.nasa.gov/. Accessed 20 Oct 2022.
69. A. Eldering, T.E. Taylor, C.W.O. Dell, R. Pavlick, *The OCO-3 Mission; Measurement Objectives and Expected Performance Based on One Year of Simulated Data*, vol. 3, no. June 2010, 2018.
70. D. Wu et al., A Lagrangian approach towards extracting signals of urban CO2 emissions from satellite observations of atmospheric column CO2 (XCO2): X-Stochastic Time-Inverted Lagrangian Transport model ('X-STILT v1'). Geosci. Model Dev. **11**(12), 4843–4871 (2018). https://doi.org/10.5194/gmd-11-4843-2018
71. J. Stoller-Conrad, Integrating Carbon from the ground up: TCCON turns ten. Earth Obs. **26**(4), 13–17 (2014)
72. D. Wunch et al., Calibration of the Total Carbon Column Observing Network using aircraft profile data. Atmos. Meas. Tech. **3**(5), 1351–1362 (2010)
73. Toon, "Toon, G., Blavier, J.-F., Washenfelder, R., Wunch, D., Keppel-Aleks, G., Wennberg, P., Connor, B., Sherlock, V., Griffith, D., Deutscher, N., Notholt, J., 2009. Total column carbon observing network (TCCON). In: Advances in Imaging. Optical Society of Ame," p. 2009, 2009.
74. Y. Yuan, R. Sussmann, M. Rettinger, L. Ries, H. Petermeier, A. Menzel, Comparison of continuous in-situ CO2 measurements with co-located column-averaged XCO2 TCCON/satellite observations and carbontracker model over the Zugspitze region. Remote Sens. **11**(24), 2981 (2019)
75. S. Oshchepkov et al., Simultaneous retrieval of atmospheric CO2 and light path modification from space-based spectroscopic observations of greenhouse gases: Methodology and application to GOSAT measurements over TCCON sites. Appl. Opt. **52**(6), 1339–1350 (2013). https://doi.org/10.1364/AO.52.001339

76. TCCON Data Archive, The TCCON Data Archive. https://tccondata.org/. Accessed 14 Oct 2022.
77. M. Sheng, L. Lei, Z.C. Zeng, W. Rao, H. Song, C. Wu, Global land 1° mapping dataset of XCO2 from satellite observations of GOSAT and OCO-2 from 2009 to 2020. Big Earth Data **00**(00), 1–21 (2022). https://doi.org/10.1080/20964471.2022.2033149
78. The CEOS Database, *The European Space Agency, CEOS Database*. http://database.eohandbook.com/database/missiontable.aspx. Accessed 16 Nov 2022.

Chapter 3
CO_2 Transportation Facilities: Economic Optimization Using Genetic Algorithm

Farzad Hourfar, Mohamed Mazhar Laljee, Ali Ahmadian, Hedia Fgaier, Ali Elkamel, and Yuri Leonenko

F. Hourfar (✉)
Department of Chemical and Materials Engineering, University of Alberta, Edmonton, AB, Canada

Department of Chemical Engineering, University of Waterloo, Waterloo, ON, Canada
e-mail: fhourfar@uwaterloo.ca

M. M. Laljee
Department of Chemical Engineering, Indian Institute of Technology Ropar, Rupnagar, Punjab, India
e-mail: 2019chb1051@iitrpr.ac.in

A. Ahmadian
Department of Electrical Engineering, University of Bonab, Bonab, Iran

Department of Chemical Engineering, University of Waterloo, Waterloo, ON, Canada

H. Fgaier
Department of Mathematics and Information Science, Full Sail University, Florida, USA

Valencia College, Orlando, FL, USA
e-mail: fgaier@alumni.uoguelph.ca

A. Elkamel
Department of Chemical Engineering, University of Waterloo, Waterloo, ON, Canada

Department of Chemical Engineering, Khalifa University, Abu Dhabi, United Arab Emirates
e-mail: aelkamel@uwaterloo.ca

Y. Leonenko
Department of Earth and Environmental Sciences, University of Waterloo, Waterloo, ON, Canada

Department of Geography and Environmental Management, University of Waterloo, Waterloo, ON, Canada
e-mail: leonenko@uwaterloo.ca

A. Ahmadian et al. (eds.), *Carbon Capture, Utilization, and Storage Technologies*, Green Energy and Technology, https://doi.org/10.1007/978-3-031-46590-1_3

3.1 Introduction

According to recent studies, it has been proven that reducing greenhouse gas (GHG) emissions is imperative to prevent global warming and protect the environment [1, 2]. One of the viable options of GHG reduction is carbon capture and storage (CCS) technologies [3] in which CO_2 is captured from different sources (such as power plants), then it is transported through pipelines [4, 5], and finally it is being sequestrated for long term in appropriate onshore/offshore reservoirs to prevent entering the atmosphere, which results in reducing adverse greenhouse gas impacts. In recent decades, despite the advancement of carbon capture technology to the point of commercial deployment and acknowledgement of underground reservoir storage as a secure solution, CO_2 transportation systems are still a challenging issue [6]. The most expensive components of a CCS chain are the CO_2 capture technologies [7]. However, optimal CO_2 transportation facility design can drastically lower the project's overall cost [8–11], especially when the source-sink distance is greater than 100 km. Recent studies demonstrate that the cost of transport facilities in a CCS project is more than anticipated [12]. Therefore, they must be designed in an economically optical manner. Moreover, a cost model intertwined with the pipeline's hydrodynamic model is necessary [13].

Several cost models for calculating the capital cost of CO_2 pipelines are available in the literature. For example, a linear cost model with correction parameters to account for terrain and region variability has been presented in [14]. Gao et al. in [15] introduced a cost model that accounts for pipeline weight which is calculated using its length, thickness, and diameter. In [16], McCoy and Rubin developed a cost model which incorporates investments for material, labor, and right of way. In addition, many cost models have been developed based on the pipeline capacity (flowrate of the CO_2 [17–19]). However, Knoope et al. in [20] have shown that only those models which consider material weight account for the cost of steel. In this chapter we apply a weight-based model for estimating pipeline capital costs.

Literature contains a variety of hydrodynamic models to determine the diameter of pipelines and calculate pressure drops along them. For example, Broek et al. in [21] have developed a hydraulic equation to estimate the optimal diameter of pipelines based on their length, mass flowrate, and fanning friction factor [22]. McCoy and Rubin in [16] have utilized a model that incorporates average values of fluid temperature and compressibility along the pipeline. Due to the correlation between pipeline diameter and the friction factor, iterative techniques are needed to calculate the diameter. The results utilizing PIPESIM software to simulate CO_2 pipes show that McCoy and Rubin's model is highly accurate [20]. In most cost models corresponding to long distance CO_2 transportation, different operational constraints are assumed, for example, booster stations at fixed locations with fixed powers. In this regard, Wildenborg et al. in [23] suggested placing one booster station every 200 km, whereas Heddle et al. in [14] proposed boosters at intervals of 100 km. Dongjie et al. in [24] added fixed capacity boosters whenever the pressure fell under 10 MPa.

This study discusses a much more general framework which allows different capacities for the booster stations. Furthermore, the boosters can be installed at any location, depending on pressure losses along the pipeline.

Managing the hydrodynamic constraints and the cost function's high nonlinearity at the same time is required to reduce the overall cost of the transportation infrastructure. For CCS infrastructure optimization involving nonlinear cost functions, researchers have suggested optimization techniques such as mixed integer programming and greedy-type algorithms. [25, 26]. Nevertheless, the search mostly terminates at local optimums. On the other hand, metaheuristic techniques are generally known as powerful optimization techniques, which can converge to the global optimum. For example, genetic algorithm (GA) is an evolutionary method inspired by natural evolution that finds use in a wide range of engineering problems. Traditional derivative-based systems determine each subsequent step in the optimization process using a deterministic calculation, whereas GA chooses a population of randomly generated solutions that evolve toward the optimum. In GA terminology, each population member that corresponds to a feasible solution is called a chromosome. Sequential steps taken by chromosomes to evolve are called generations. In every generation, an objective or fitness function evaluates the chromosomes. In order to create the set of new chromosomes for the future generation (offspring), either separate chromosomes from the current generation are combined (using the crossover operator) or an existing chromosome is changed (using the mutation operator). To introduce the next generation with constant population size, parents and offspring are opted based on the value of fitness function. Generally, GA converges to the optimal solution after several generations, while using suitable chromosomes [27, 28]. Numerous problems, including flow shop scheduling [29], machine learning tasks [30], optimum gas transmission design [31], and optimum water distribution system design [32–35], have been successfully solved using this algorithm.

In this chapter, a pipeline material weight-based cost model has been considered to include the steel price. In addition, variable locations and capacities of booster stations that compensate for pressure losses in the pipeline are assumed in the design procedure. Moreover, the Peng-Robinson equation of state (EoS) is used to reflect the variations in CO_2 density caused by pressure changes along the transportation system. For design flexibility and reduction of total cost, all pipe segments are free to have different diameters. Further, using GA optimization method enhances the probability of finding the global optimum. Different equations of state can replace Peng-Robinson EoS, without substantial changes to the algorithm. Finally, the effects of topographical factors on CO_2 pressure drop and the corresponding optimal solution are presented.

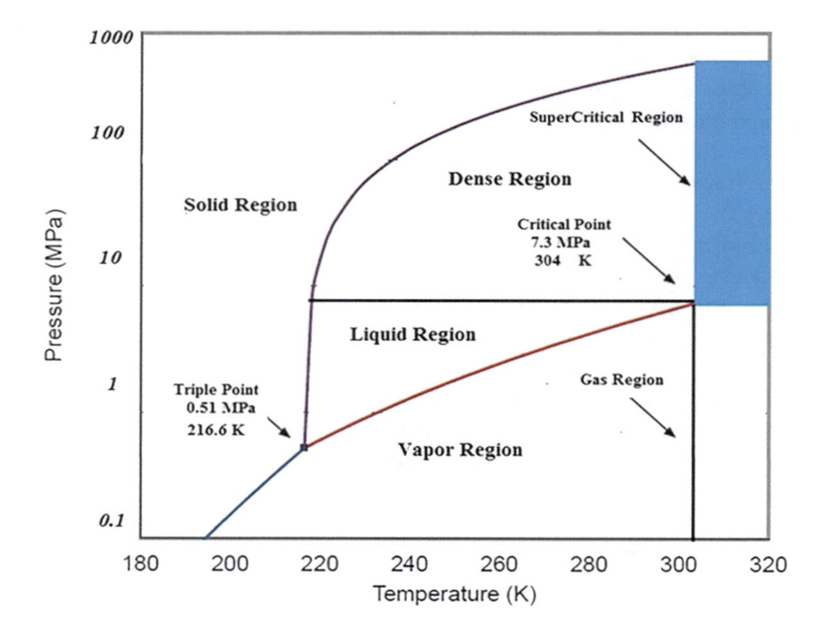

Fig. 3.1 Pure CO_2 phase diagram

3.2 Problem Statement

3.2.1 Hydrodynamic Model

It is possible to transport CO_2 in the form of liquid, gas, two-phase fluid, or supercritical fluid. As the critical point of CO_2 is around the operational temperature, any small variations in pressure and temperature can cause significant changes in CO_2 physical properties (such as phase, density, and compressibility). To prevent operational problems such as corrosion and cavitation, CO_2 is generally transported in the form of supercritical or dense fluid [36]. By considering a pure CO_2 phase diagram (Fig. 3.1), the supercritical phase is defined to be at pressures above 7.34 MPa, and temperatures above 304.35 K. The region of dense fluid is defined at temperatures less than the critical temperature, but at pressures greater than the critical pressure. To prevent the problems of two-phase flow during the transportation process, minimum operational pressures between 7 and 10 MPa have been proposed in the literature [14, 16, 23, 37]. More studies have also confirmed that the minimum pressure along the pipeline should be greater than the critical pressure to guarantee single-phase flow [38–40]. The high density of CO_2 in dense phase or supercritical phase, which helps increase the efficiency of the transportation process.

In this study, we investigate different scenarios for a trunk pipeline connecting a source and sink separated by a distance of 1000 km, which transports CO_2 at a mass flowrate of 750 kg/s. In general, coal-based power plants are among the largest source of CO_2 emitters. A 500 MW coal power plant typically generates around 2–3 million tons of CO_2 per year [16]. Approximately 200 coal-based power plants with net capacity above 2000 MW are now under operation. The capacity of about 11 of them are greater than 4500 MW [41, 42]. In this study, we focus on a mass flowrate of 750 kg/s from source to sink, which is around 23.7 million tons of CO_2 transportation per year. This amount corresponds to the annual CO_2 emissions from a 4000 MW coal-based power plant. The ability of a pipeline to transport materials is influenced by various elements, including the permissible delivery pressure range, choice of route, and the inherent characteristics of the substance being transported. Within this section, our focus will be on land-based pipelines, thereby sidestepping complications associated with placing pipelines on the ocean floor, which arise due to factors like buoyancy impacts. As materials are transported through the pipeline, declines in pressure occur due to both frictional effects and alterations in elevation. Consequently, the inclusion of booster stations becomes imperative to re-elevate the pressure of CO2, ensuring its maintenance in a state of supercriticality or as a dense liquid.

As a broad structure, the pipeline sections connecting individual booster stations have the potential to vary in terms of both diameter and length. The reduction in pressure within each section is determined through computations involving mass and momentum equilibrium equations. The mass equilibrium equation is formulated for the two termini of a given pipeline segment, namely, between point i and point $i + 1$:

$$\dot{m}_i = \dot{m}_{i+1} \rightarrow \rho_i \left(\pi D_i^2/4\right) \overline{v}_i = \rho_{i+1} \left(\pi D_i^2/4\right) \overline{v}_{i+1} \rightarrow \rho_i \overline{v}_i = \rho_{i+1} \overline{v}_{i+1} \tag{3.1}$$

$\dot{m}_i$ *and* $\dot{m}_{i+1}$ are the mass flowrates at points i and $i + 1$ in *kg/s*. ρ_i and ρ_{i+1} are the densities of CO_2 at points i and $i + 1$ in kg/m^3. $\overline{v}_i$ and $\overline{v}_{i+1}$ denote the average velocities of CO_2 at those points in *m/s*. Finally, D_i is the diameter of the pipeline in that segment, in meter *(m)*.

The following practical assumptions are also considered: (i) The operational temperature is less than the critical point (ambient temperature is assumed); (ii) the pressure is higher than the critical pressure to avoid formation of two-phase flows but not so high as to threaten pipeline integrity. So, typical CO_2 pipeline operating pressures and temperatures are in the ranges of 10 to 15 MPa and 15 to 30 °C, respectively, which is similar to [43, 44].

Meanwhile, Peng-Robinson (PR) equation of state (EOS) is used to calculate the density:

$$P = \frac{RT}{v - b} - \frac{a\alpha(T)}{v^2 + 2bv - b^2}, \tag{3.2}$$

in which

$$a = \frac{0.45724 R^2 T_c^2}{P_c}, \tag{3.3}$$

$$b = \frac{0.07780 R T_c}{P_c}, \tag{3.4}$$

and,

$$\alpha(T) = \left[1 + \left(0.37464 + 1.54226\omega - 0.26992\omega^2\right)\left(1 - \sqrt{T_r}\right)\right]^2 \tag{3.5}$$

T (K) and P (MPa) are CO_2 temperature and pressure, respectively. In these equations, $T_c = 304.35\ K$, $P_c = 7.34\ MPa$, and $\omega = 0.2236$ are the values of critical temperature, critical pressure, and acentric factor of pure CO_2, respectively. In addition, T_r is the reduced temperature defined as T/T_c and ν (m^3/mol) is the molar volume.

Generally, "Groupe Européen de Recherches Gazières" (GERG) is one of the most accurate EOS with the absolute average deviation (AAD) of 1%, while PR has an AAD of 4.7%. However, within the selected operational condition, PR achieves an AAD of 2.1% [44], which is suitable for most practical applications. It should be noted that the total accuracy can be improved by including the CO_2 volume correction term in PR EOS, at the expense of additional computational costs. Peneloux shift parameter helps to have better density estimation, especially in dense regions with pressures above 20 MPa, but for the pressure range of 10–15 MPa, there is no significant improvement in the accuracy.

The energy equilibrium (expressed in terms of head) is established between two specific points (i and $i + 1$) within a segment of the pipeline through the subsequent formulation:

$$\frac{p_i}{\rho_i g} + \frac{\overline{v}_i^2}{2g} + z_i = \frac{p_{i+1}}{\rho_{i+1} g} + \frac{\overline{v}_{i+1}^2}{2g} + z_{i+1} + h_{f_i} \tag{3.6}$$

where p_i, p_{i+1} (Pa) are the pressures at points i and $i + 1$, z_i and z_{i+1} (m) denote the elevations at points i and $i + 1$, and h_{fi} (m) is frictional head loss. Other types of pressure loss due to either changes in pipeline geometry (e.g., bending and elbows, contraction/expansion, etc.) or the presence of different piping components (e.g., valves, pipe connections, and fittings) can be considered. These losses are known as non-frictional or minor losses which can be added to Eq. 6 as a sum of head losses for each source $h_m = K\ v^2/2g$, where K is the loss coefficient and v is the flow velocity. For-long distance pipelines, minor losses are negligible in comparison with frictional losses.

The frictional head loss of the pipe segment i is derived as:

$$h_{f_i} = \frac{2 f_F \overline{v}^2 L_i}{g D_i} \tag{3.7}$$

$\overline{\upsilon}$ (m/s) is the average velocity of CO_2 in the ith segment, and f_F is the fanning friction factor [21], which is obtained through the following equation for the Reynolds number range of $4000 < Re < 4 \times 10^8$, and for relative roughness range of $10^{-7} < \varepsilon/D < 0.05$ [22]:

$$\frac{1}{\sqrt{f_F}} = -4\log_{10}\left\{\frac{\varepsilon}{3.7D} - \frac{5.02}{\mathrm{Re}}\log_{10}\left[\frac{\varepsilon}{3.7D} - \frac{5.02}{\mathrm{Re}}\log_{10}\left(\frac{\varepsilon}{3.7D} + \frac{13}{\mathrm{Re}}\right)\right]\right\} \tag{3.8}$$

ε (*m*) is the roughness of the pipe wall, and *Re* is the Reynolds number which is defined as:

$$\mathrm{Re} = \frac{4\dot{m}}{\mu\pi D}, \tag{3.9}$$

where μ (*kg/m.s*) is the absolute viscosity. The average velocity is $\bar{v} = (v_i + v_{i+1})/2$, and the velocity at each point can be calculated from:

$$\overline{v} = \frac{4\dot{m}}{\pi\rho D^2} \tag{3.10}$$

Densities ρ_i and ρ_{i+1} are used in points i and $i + 1$ to calculate v_i and v_{i+1}.

In this context, viscosity is considered to remain constant and is set at the mean value aligning with the prevailing thermodynamic conditions in operation. However, it is important to note that as pressure fluctuates along the pipeline, the viscosity also undergoes alterations. The utmost disparity between viscosity and its average value remains below 10% [45]. This amount of deviation causes slight changes (less than 0.5%) in frictional factor given by Eq. 3.8.

To take into account the impacts of different slopes of the terrain, the variation in the elevation of a pipe ($z_{i+i} - z_i$) is calculated as a function of the segment length, L_i, the angle with the horizontal line θ_i, using the following trigonometric formula:

$$z_{i+1} - z_i = L_i \sin\theta_i \tag{3.11}$$

Replacing derived variables from Eqs. (3.1), (3.7), and (3.11), into Eq. (2.6) results in:

$$p_{i+1} = p_i\left(\frac{\rho_{i+1}}{\rho_i}\right) + \frac{1}{2}\rho_{i+1}\overline{v}_i^2\left(1 - \frac{\rho_i^2}{\rho_{i+1}^2}\right) - \rho_{i+1}gL_i\sin\theta_i - \left(\frac{2f_{F_i}\overline{v}_i{}^2L_i}{gD_i}\right)\rho_{i+1}g \tag{3.12}$$

Given the interrelation between carbon dioxide's density and pressure, the Peng-Robinson equation of state (3.2) is employed. This equation facilitates the computation of density (and consequently pressure) at the termination of each pipeline segment, employing an iterative methodology. Using Eq. (3.12), the pressure at the end of each segment (p_{i+1}) is determined, given the pressure value at the beginning of each pipe segment (p_i). In Table 3.1, the parameters of the hydrodynamic model of CO_2 are presented.

Table 3.1 CO_2 hydrodynamic model parameters

Parameter	Physical meaning
p	CO_2 pressure
ρ	CO_2 density
ν	CO_2 flow velocity
m˙	CO_2 mass flowrate
D	Pipeline internal diameter
f_F	Fanning frication factor
Re	Reynolds number
ε	Pipeline wall roughness
θ	Angle between pipeline and horizon
z	Pipeline elevation
L	Pipeline length

3.2.2 *Cost Model for CO_2 Transportation*

The total cost for CO_2 transportation process ($Cost_{total}$) consists of the capital cost ($Cost_{capital}$), the operation and maintenance cost ($Cost_{O\&M}$), and the energy consumption cost ($Cost_{energy}$):

$$Cost_{Total} = Cost_{capital} + Cost_{O\&M} + Cost_{energy}. \tag{3.13}$$

Evidently, the comprehensive structure of the total cost equation can be extended by incorporating additional influential elements such as: (1) compensation for rights of way and (2) expenses related to the personnel engaged in engineering and oversight tasks. However, in this chapter we will restrict focus to Eq. (3.13).

3.2.2.1 Capital Cost

The capital cost for the transportation system is divided into two parts: (1) cost of the pipe (C_{pipe}) and (2) cost of the booster stations ($C_{booster}$).

$$Cost_{capital} = Cost_{pipe} + Cost_{booster}. \tag{3.14}$$

The capital cost for the pipe is a function of pipe diameter and thickness, length, and the used materials. For CO_2 transportation, usually X70 steel is used [13], whose wall roughness (ε) is 45.7 μm. In this chapter, the price for X70 steel is assumed to be 1025 *USD/ton*.

Hence, the material cost for the pipeline is obtained as:

$$cost_{pipe} = 1.025\pi\rho_p L \frac{\left(D + 2t_p\right)^2 - D^2}{4} \tag{3.15}$$

in which ρ_p (*kg/m³*) and t_p (*m*) are the density and the wall thickness of the pipeline, respectively. Clearly, Eq. (3.15) does not include the manufacturing cost. Moreover, according to McCoy and Rubin [16], the minimum pipeline thickness (t_p) is obtained based on the maximum operating pressure as:

$$t_p = \frac{P_{\max} D}{2SFE}, \tag{3.16}$$

where $P_{\max}$ denotes maximum operating pressure (e.g., 15 *MPa*), S is the minimum yield stress (e.g., 483 *MPa*), F is the design factor (e.g., 0.72), and E is the longitudinal joint factor (e.g., 1.0) [16].

Conversely, the capital outlay for booster stations is computed considering both the necessary power of the pumps and the topographical features. For the sake of simplicity and without sacrificing comprehensiveness, the assumption is made that each booster station comprises solely one pump unit. Based on the cost equations introduced by McCollum and Ogden [19], the capital cost (*USD*) is calculated as:

$$\mathrm{C}_{\text{booster}} = F_L \times (7.82 W_P + 0.46) \times 10^6 \tag{3.17}$$

in which W_P (*MW*) denotes the pump capacity, calculated by input pressure and output pressure of the pump [21]:

$$W_p = \frac{\dot{m}}{\rho} \frac{(P_{\text{out}} - P_{\text{in}})}{\eta_{\text{booster}}} \tag{3.18}$$

where $\dot{m}$ (*kg/s*) is the mass flowrate, ρ (*kg/m³*) denotes the density of pumping fluid, P_{out} (*MPa*) is the pump's outlet pressure, and P_{in} (*MPa*) is the inlet pressure of the pump. Moreover, in Eq. (17) F_L denotes the location factor which changes geographically. For example, for North American terrain, $F_L = 1$ is an acceptable assumption [19].

3.2.2.2 Operation and Maintenance Cost

The operation and maintenance (O&M) cost for CO_2 transportation facilities generally includes two significant portions: (1) the pipeline portion ($\mathrm{O\&M_{pipe}}$) and (2) the pump portion ($\mathrm{O\&M_{pump}}$):

$$\mathrm{Cost_{O\&M}} = \mathrm{O\&M_{pipe}} + \mathrm{O\&M_{pump}} \tag{3.19}$$

McCollum and Ogden [19] have reported that the annual operation and maintenance cost for a CO_2 pipeline (*USD/year*) is obtained by (after adjusting to SI units):

$$\begin{aligned} \text{O\&M}_{\text{pipe}} = 120000 + 0.61(913898D + 0.899L - 259269) \\ + 0.7(1547440D + 1.694L - 351355) + 24000, \end{aligned} \tag{3.20}$$

where D (m) is the internal diameter of the pipeline and L (km) denotes its length.

The operation and maintenance cost of a pump is a function of nominal power. For example, Eq. (3.21) is suggested in [19] to calculate $O\&M_{pump}$ in *USD* for W_p less than *2 MW*:

$$\text{O\&M}_{\text{pump}} = -176.864W_p^2 + 671.665W_p + 159.292, \tag{3.21}$$

where W_p in (*MW*) denotes the pump power. However, for the sake of simplicity, to allow booster pumps of larger capacity and for not imposing any additional inequality constraint to the optimization problem (e.g., $W_p < 2$ MW), operation and maintenance cost of the booster pumps is estimated as a specified percentage of their capital costs (generally between 1.5% and 5%). We assume an average value of 3.25%.

3.2.2.3 Energy Consumption Cost

Energy is generally consumed in the form of electricity in the CO_2 transportation facilities by the booster pumps. The cost of the consumed electrical energy is related to the pump capacity and as [19]:

$$\text{Cost}_{\text{energy}} = COE \times \left(\sum_{i=1}^{N} W_{P_i} \times CF_i \right) \times 8760, \tag{3.22}$$

in which N is the number of active pumps, CF denotes the capacity factor of the pumps (e.g., 0.8), and COE is the electricity price (e.g., 20 *\$/MWh*).

3.2.3 *Optimization Problem*

The primary objective in this context is to minimize the overall cost of transporting CO2 over a predetermined distance between the source and the destination. To keep the pressure above a certain value to ensure CO_2 remains in a super-critical condition, booster stations may be required to be installed along the pipeline. In this chapter, we assume the pipeline inlet pressure at the source is 15 MPa, and the pipeline outlet pressure at the sink is 10 MPa. According to Eq. (3.13), to minimize the total cost of the transportation facilities, the optimal values of the following variables should be obtained:

- Number of required booster stations.
- Power of booster stations.
- Length of pipe segments.
- Diameter of pipe segments.

Meanwhile, the following practical constraints are considered in this optimization problem:

- Each booster station increases the pressure of CO_2. So, for booster station i, the discharge pressure, P_{d_i}, should be more than the suction pressure, P_{s_i}:

$$\frac{P_{d_i}}{P_{s_i}} \geq 1 \; . \tag{3.23}$$

- The pressure should always be less than the maximum allowed pressure for any segment of diameter d_i, to maintain the mechanical integrity of segment i:

$$P_{d_i}^{\min} \leq P_{d_i} \leq P_{d_i}^{\max}, \tag{3.24}$$

in which $P_{d_i}^{\max}$ and $P_{d_i}^{\min}$ denote the maximum and minimum allowed pressure for segment i.

- To increase the efficiency in CO_2 transportation process, a dense liquid or supercritical state is necessary. So, the CO_2 pressure should be more than a specified value:

$$P_{s_i} \geq P_{s_i}^{\min} P_{d_i}^{\min} \leq P_{d_i} \leq P_{d_i}^{\max}. \tag{3.25}$$

- Maximum power of the pump is obtained based on the values of $P_{d_i}^{\max}$ and $P_{d_i}^{\min}$, as follows:

$$W_p \leq \frac{\dot{m}}{\rho} \frac{\left(P_{d_i}^{\max} - P_{d_i}^{\min}\right)}{\eta_{\text{booster}}}. \tag{3.26}$$

- Furthermore, the lengths of pipe segments need to be greater than a certain value which economically justifies the placement of booster stations. Meanwhile, the pipeline diameter in each segment is necessary to be in a standard piping range to guarantee the existence of the size in the market:

$$L_i \geq L_i^{\min} \tag{3.27}$$

$$D_i^{\min} \le D_i \le D_i^{\max} \tag{3.28}$$

In addition to the inequality constraints, two equality constraints are as follows:

- The total distance between the source and the destination is equivalent to the sum of the individual lengths of the pipeline segments:

$$L_t = \sum\nolimits_{i=1}^{N+1} L_i, \tag{3.29}$$

in which N is the number of boosters, and L_t is the total pipe length, which depends on the source-sink distance and pipeline configurations.

- The input pressure of each booster station is calculated by the discharge pressure of the previous booster station and the value of pressure drop in the previous segment, considering the hydrodynamic model such as Eq. (3.12):

$$P_{s_{i+1}} = P_{d_i}\left(\frac{\rho_{i+1}}{\rho_i}\right) + \frac{1}{2}\rho_{i+1}\overline{v}_i^2\left(1 - \frac{\rho_i^2}{\rho_{i+1}^2}\right) - \rho_{i+1} g L_i \sin\theta_i - \left(\frac{2 f_{F_i} \overline{v}_i^{\,2} L_i}{g D_i}\right)\rho_{i+1} g \tag{3.30}$$

Based on the above explanations, the defined CO_2 transportation optimization problem is straightforwardly formulated in the general form as:

$$\begin{aligned} &\min f(x) \\ &\text{subject to } g_i(x) \le 0 \text{ for each } i \in \{1, \cdots, m\}, \\ &\qquad h_j(x) = 0 \text{ for each } j \in \{1, \cdots, p\}, \\ &\qquad x \in X, \end{aligned} \tag{3.31}$$

where X is a subset of R^n. Moreover, f, g_i, and h_i are real-valued functions on X for each i and each j. In addition, m, n, and p are positive integers. In a nonlinear optimization problem, at least one of f, g_i, and h_i is nonlinear. Clearly, f, g_i, and h_i are easily obtained, using the highlighted equations and inequalities in Sects. 3.2.2 and 3.2.3.

3.3 Results and Discussion

Figure 3.2 demonstrates the considered optimization problem and the values of defined parameters are presented in Table 3.2. we have assumed that pure CO_2 is transported through the pipeline. Moreover, the impact of corrosion on the pipeline cost is negligible.

In Sect. 3.3.1, the effect of pipeline characteristics on the value of pipeline pressure drop is covered, and in Sect. 3.3.2, the total cost's sensitivity to (1) pipeline diameter and (2) booster station power is investigated. The constrained optimization routine established in Sect. 3.2 is then applied to two case studies in Sect. 3.3.3 to find the optimal design parameters that minimize the pipeline's total cost. To this end, the *ga function* in Matlab is used to implement the GA algorithm for finding optimal results subject to equality, inequality, and nonlinear constraints explained in Sect. 3.2.3.

Parameters of the Matlab *ga function* are tuned as:

- Termination tolerance: 10^{-6}.
- Initial population: *50*.

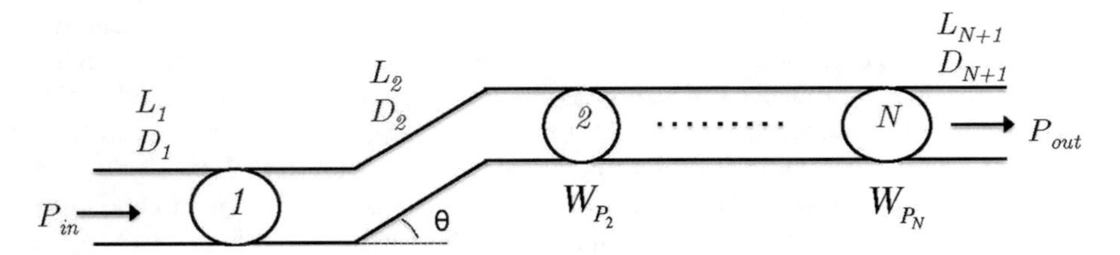

Fig. 3.2 Schematic of design parameters in pipeline segments

Table 3.2 Optimization problem parameters

Parameter	Description	Value
L	Total pipeline length	1000 km (for horizontal pipe)
T	Temperature	298.15 K
μ	CO_2 viscosity	82.6E-6 Pa.s
ε	Pipeline wall roughness	45.7E-6 m
P_{in}	Inlet pressure (at pipeline's source)	15 MPa
P_{out}	Outlet pressure (at pipeline's sink)	10 Mpa
ρ_P	Pipeline material (X70 steel) density	7700 kgm^{-3}
$P_{s_i}^{max}$	Minimum operating and minimum suction pressure for pipe segment *I* and pump *i*, respectively	9 MPa
$P_{d_i}^{max}$	Maximum operating pressure for pipe segment *i*	15 MPa
L_i^{min}	Each pipeline segment's minimum length	100 km
D_i^{min}	Each pipeline segment's minimum diameter	0.6 m
D_i^{max}	Each pipeline segment's maximum diameter	1 m

Moreover, based on presented explanations and formulae in Sects. 3.2.2. and 3.2.3, the objective/fitness function and the constraints (equality and inequality, linear and/or nonlinear) are defined. [46, 47] provide additional details on tuning parameter values, theoretical underpinnings, and implementation of GA-based optimization techniques in Matlab. It should be highlighted that compared with traditional optimization methods, using GA enhances the chance of converging to the global optimum. Alternatively put, running the simulations with various initial values makes finding the optimal solution substantially more probable. Clearly, to apply this strategy when the size of the optimization problem (number of decision variables) is large, a high computational power is required. For instance, despite the considered problem's dimensionality being not too large, the average run time for each simulation is around 4–5 hours.

3.3.1 Correlation Between Design Parameters and Pressure Drop

In this section, we consider a trunk pipeline transporting CO_2 at a mass flowrate of 750 kg/s in a dense liquid state, and the inlet pressure is fixed at 15 MPa. For the first analysis, the internal diameter of a 100 km long horizontal pipeline is varied from 0.6 to 1 m to study its influence on the outlet pressure. Next, the diameter is fixed at 1 m and the length is varied from 0 to 1000 km. Afterward, the slope of a 1 km long pipeline having the diameter of 1 m is varied from -60 to 60 degree. Lastly, the flowrate through a 100 km long pipe with diameter 1 m is varied from 0 to 800 kg/s. Figure 3.3 shows how the changes in the length, diameter, elevation, and CO_2 mass flowrate affect the pressure drop (and hence pressure at the pipe's outlet).

According to Fig. 3.3a, pipeline diameter changes significantly influence the pressure drop. Increasing the pipeline diameter from 0.6 m to 1 m reduces the pressure drop by about 8 MPa for a distance of 100 km. In Fig. 3.3b, it can be seen that in a 1000 km pipeline whose diameter is 1 m, CO_2 pressure almost linearly decreases by about 6 MPa from source to sink. This is roughly equal to 1.2 MPa for every 200 km. According to Fig. 3.3c, the elevation change (slope variation) of the pipeline has a profound impact on the pressure variation. For instance, for a 1 km pipeline having a diameter of 1 m, an uphill slope of 60° results in an approximate 7 MPa drop in pressure. On the other hand, a $-$ 60° downhill angle causes the pressure to increase by about 7 MPa. Lastly, Fig. 3.3d demonstrates how increasing the CO_2 mass flowrate augments the Reynolds number and reduces the CO_2 pressure at the outlet. This reduction in pressure is about 7 MPa for a 1000 km long pipeline of 1 m diameter with the mass flowrate equalling 800 kg/s.

According to Fig. 3.3, two key parameters relating to the pipeline pressure drop are the pipeline slope, and diameter. In addition, Fig. 3.3b shows that a minimum of one booster station is needed for transporting 750 kg/s of CO_2 in a 1000 km completely horizontal pipeline whose diameter is 1 m so as to ensure an outlet pressure of at least 10 MPa. Thus, in the optimization problem, the length of pipe

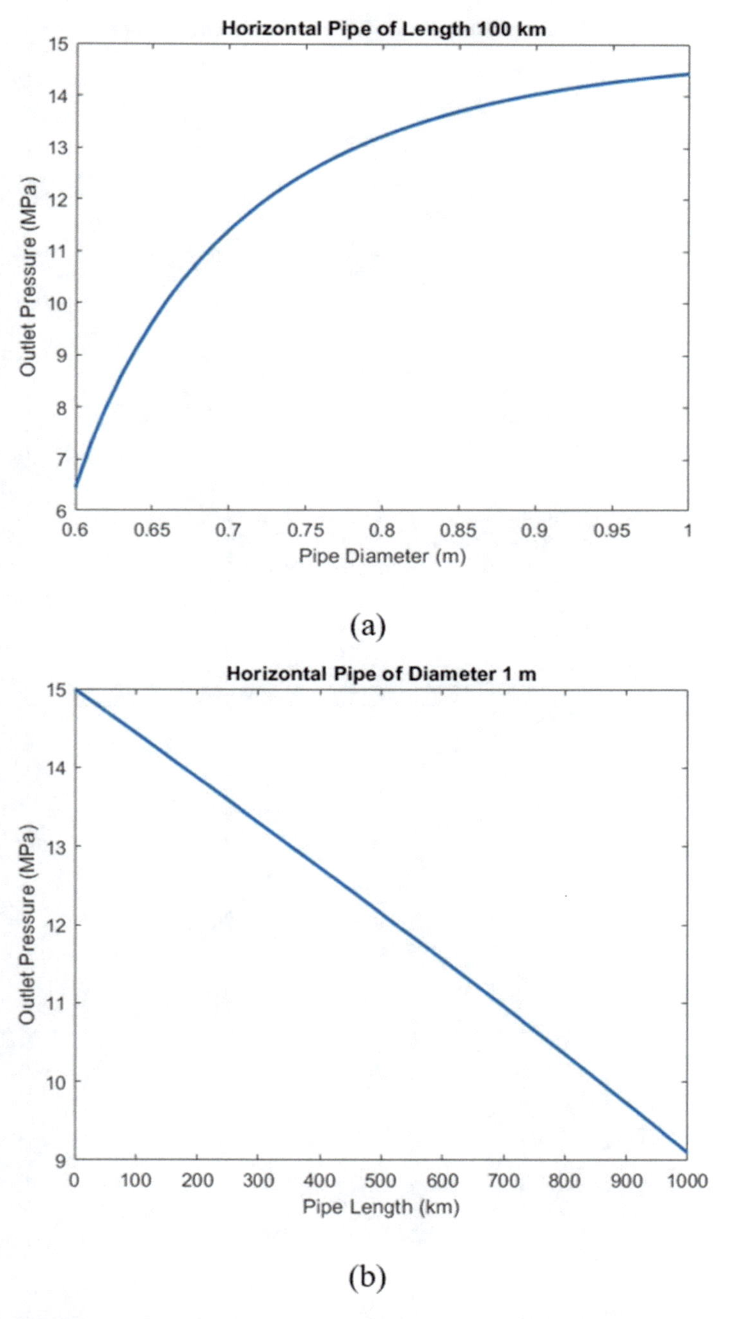

Fig. 3.3 Outlet pressure in different cases: (**a**) Diameter variations for a 100 km pipeline, (**b**) length variations for a pipeline with diameter of 1 m, (**c**) various slopes for a 1 km pipeline with diameter of 1 m, and (**d**) different mass flowrates of CO_2 for a 1000 km pipeline with diameter of 1 m

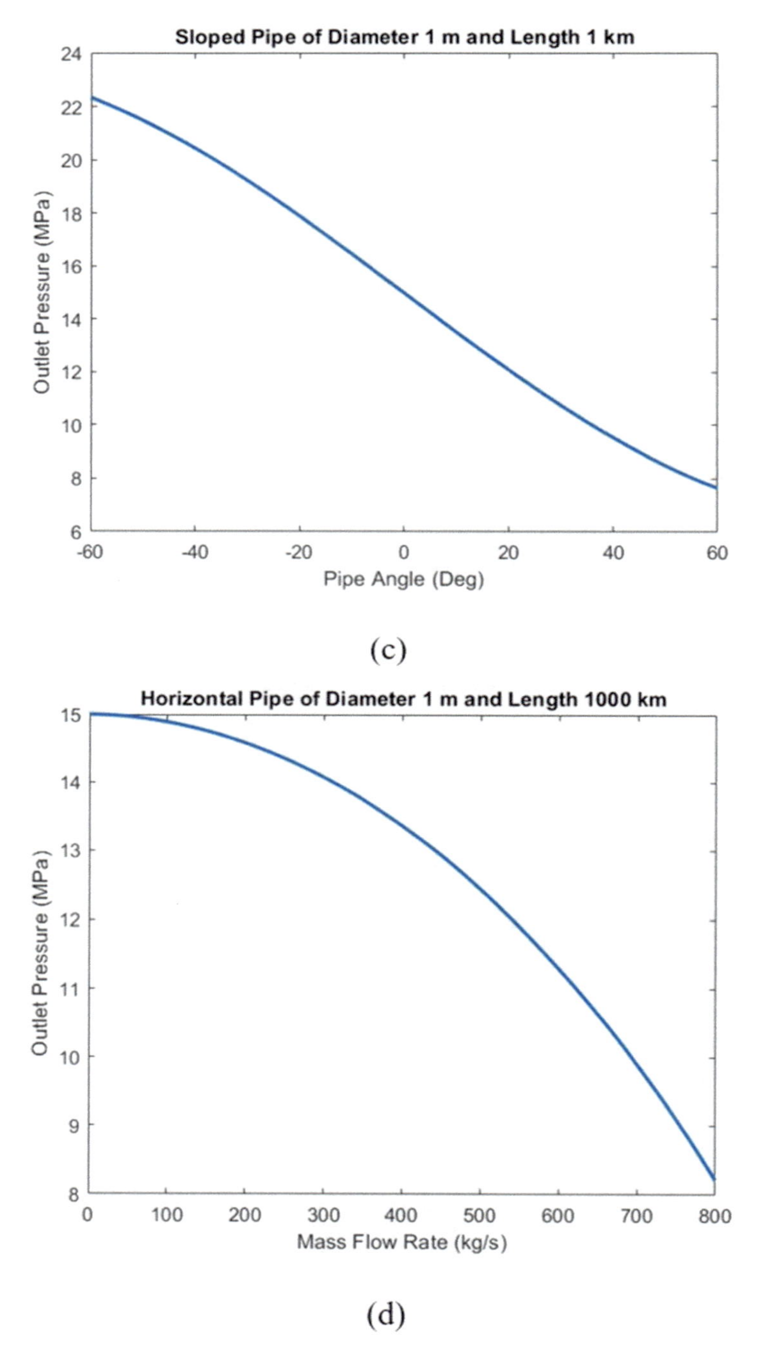

Fig. 3.3 (continued)

segments/location of booster stations and their nominal power, and the pipe segment diameters are calculated such that the pipeline's total cost is minimized.

3.3.2 Correlation Between Design Parameters and Total Cost

The sensitivity of the pipeline cost to pipeline diameter and booster station capacity is presented in Fig. 3.4. The cost of the pipe increases non-linearly with increase in diameter, while the booster station cost increases linearly with increase in power. The majority of a CO_2 transportation system's total costs are attributable to the booster stations. Hence, to compensate for pressure drops in the pipeline, the first cost-effective step is to increase the pipeline diameter instead of using booster stations.

3.3.3 Optimal Design in Selected Scenarios

The optimal transport system design is obtained for two different scenarios with a fixed CO_2 mass flowrate of 750 kg/s: (1) a horizontal trunk pipeline and (2) a trunk pipeline with uphill and downhill slopes, for an on-ground source-sink distance of 1000 km. So, the effects of mass flowrate and pipeline inclination on the pressure variation and optimal solution are demonstrated. In both cases, the pipe diameter can take on any value from 0.6 to 1 m. The inlet pressure and the maximum operating pressure are fixed at 15 MPa, while the delivery pressure at the sink must be at least 10 MPa. Lastly, the pressure in the pipe is maintained above 9 MPa (well above the critical pressure of 7.34 MPa) at all times so as to ensure a dense liquid state.

Case Study 1: Horizontal Trunk Pipeline (750 kgs^{-1})
Two different sub-cases are considered in this section. For the first one, the maximum allowable diameter of the pipeline is 0.8 m, whereas the maximum value is increased to 1 m in the second one.

In Fig. 3.5, CO_2 pressure and density profiles corresponding to optimal design for $D_i^{\max} = 0.8$ m with the number of booster stations varying from $N = 1$ to 4 can be seen. Clearly, sharp pressure peaks are seen at the location of booster stations. Moreover, in Table 3.3 the information on (1) booster power, (2) location of booster stations/length of pipe segments, and (3) pipe segment diameters is provided.

To minimize the cost, the algorithm converges to the smallest yet most cost-effective and feasible booster capacities that satisfy the constraints defined in Sect. 3.2.3. As can be clearly seen in Fig. 3.5, the CO_2 pressure is continuously greater than 9 MPa (minimum operating pressure). Furthermore, according to the presented results in Table 3.3, there is no feasible solution for the cases with one and two booster stations. Moreover, the diameters of pipeline segments tend to converge to the maximum allowable value (0.8 m) to use the maximum pipeline capacity and

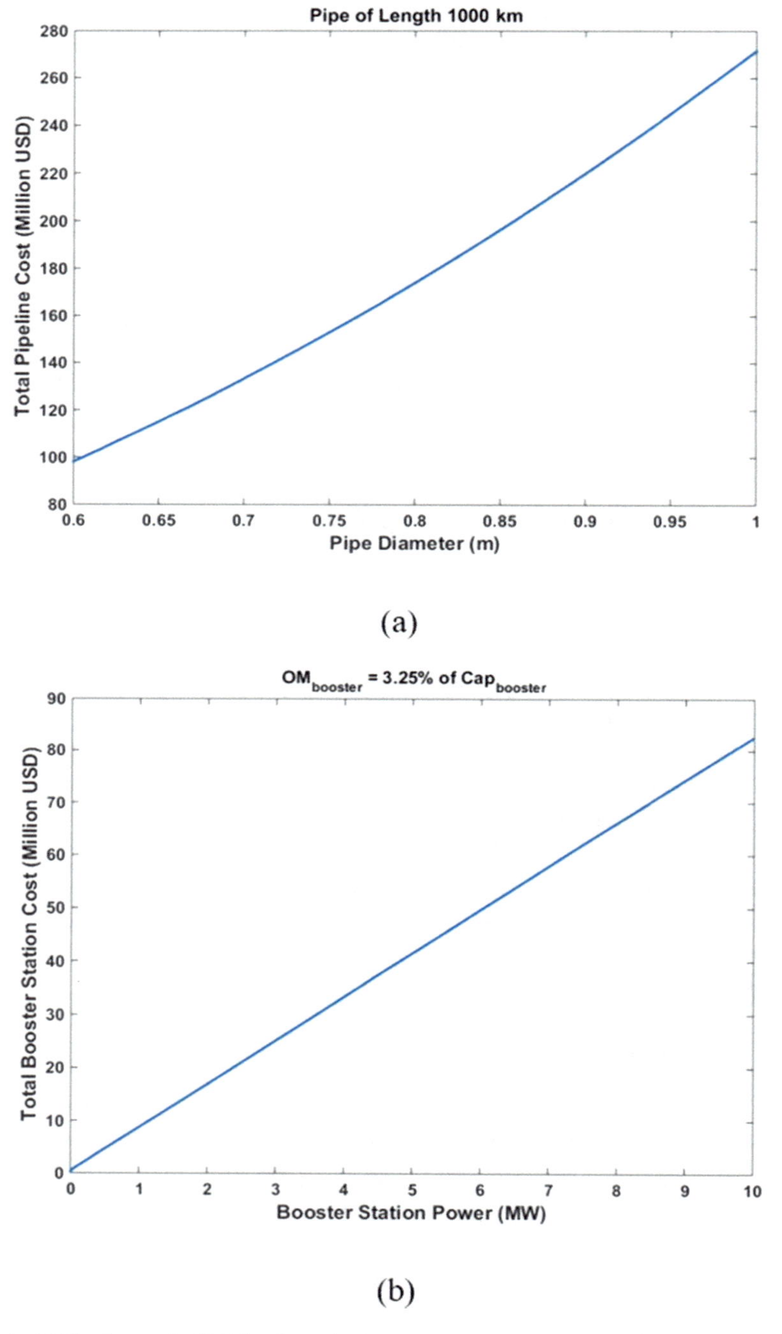

Fig. 3.4 (**a**) Total cost of a 1000 km long pipeline with different diameters; (**b**) total booster station cost for different pump powers

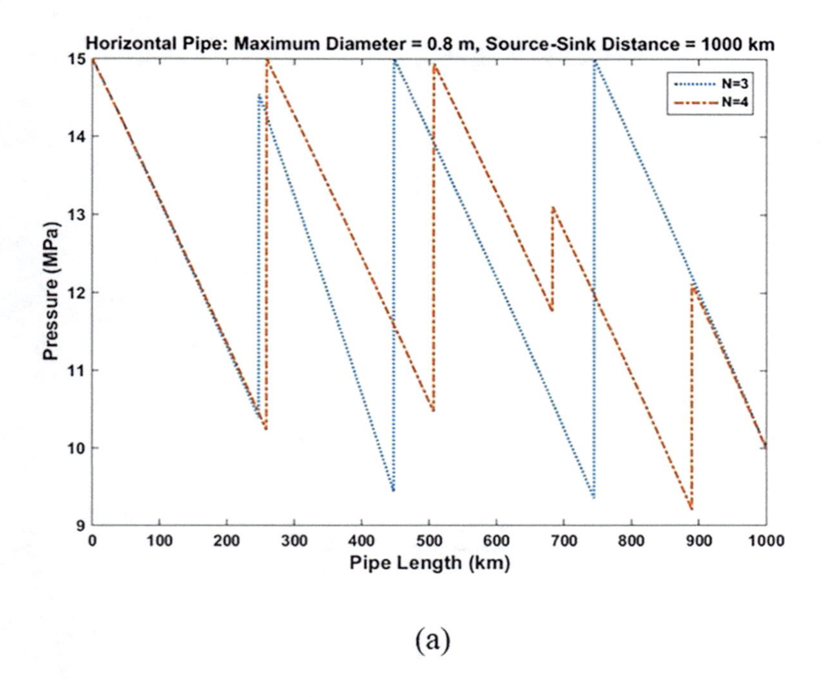

(a)

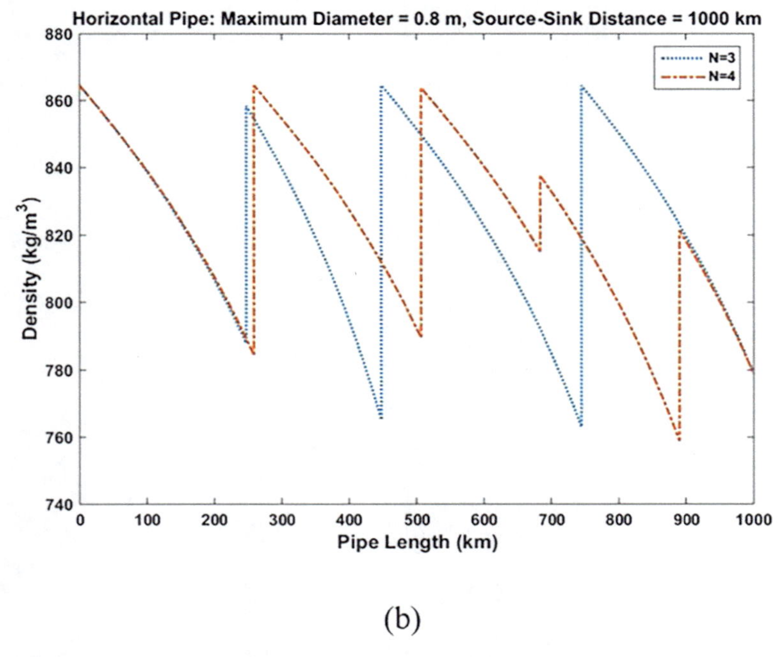

(b)

Fig. 3.5 (**a**) Pressure variation and (**b**) density variation of CO_2 for the optimal design of a 1000 km horizontal pipeline with different number of booster stations. $D_i^{max} = 0.8$ m, mass flowrate = 750 kg/s, $O\&M_{booster} = 3.25\%$ of $Cap_{booster}$

Table 3.3 Optimal design parameters for a 1000 km horizontal pipeline.$D_i^{max} = 0.8$ m, mass flowrate = 750 kg/s, $O\&M_{booster} = 3.25\%$of $Cap_{booster}$

	Booster station power (MW)				Pipeline segment diameter (m)					Pipeline segment length (km)				
Number of boosters	Wp_1	Wp_2	Wp_3	Wp_4	D_1	D_2	D_3	D_4	D_5	L_1	L_2	L_3	L_4	L_5
1	No feasible solution													
2														
3	5.267	7.266	7.390	–	0.8	0.75	0.8	0.79		247	201	297	255	
4	6.078	5.668	1.659	3.855	0.0.8	0.8	0.8	0.8	0.8	259	248	176	207	110

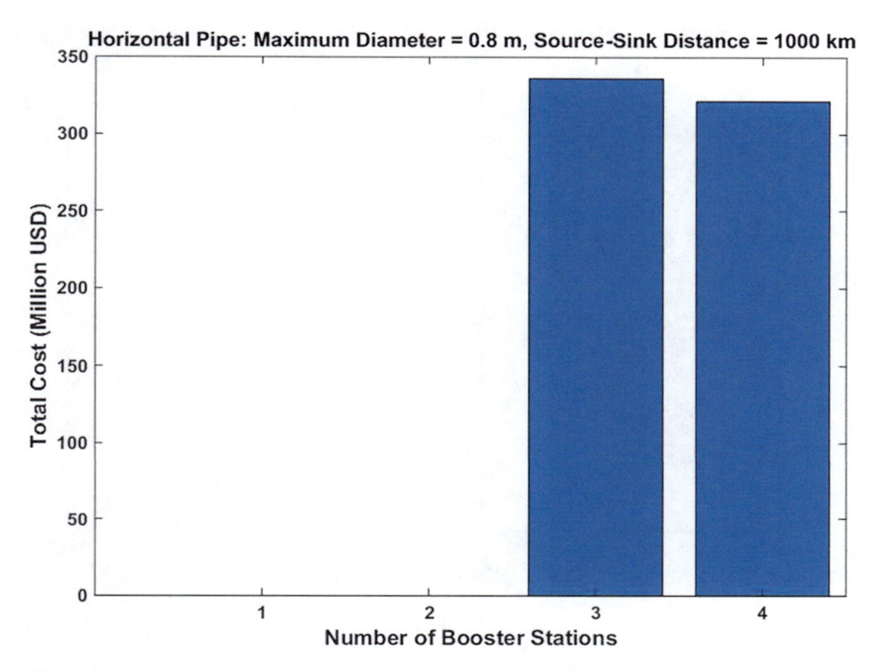

Fig. 3.6 Total cost for the optimal design of a 1000 km horizontal pipeline with different number of booster stations. $D_i^{\max} = 0.8$ m, mass flowrate = 750 kg/s, O&M$_{\text{booster}}$ = 3.25% of Cap$_{\text{booster}}$

minimize the cost of required booster stations. In other words, increasing the pipeline diameter is a more cost-effective option in comparison with increasing the power of boosting stations to maintain the pipeline pressure. Meanwhile, CO_2 density remains around 760–870 kgm^{-3} and the trend is similar to that of the pressure profile (Fig. 3.5).

In Fig. 3.6, the total costs for optimal solutions at different number of booster stations with maximum allowable diameter, $D_i^{\max} = 0.8$ m, are depicted. The algorithm cannot achieve feasible solutions for the cases with one and two booster stations due to operating constraints. It is clear that the solution with four booster stations is more cost-efficient in comparison with the three-booster-station scenario. According to available information in Table 3.3, the decrease in the total cost for the four-booster station scenario is because of the lower accumulative value of W_p in all segments. Based on Eqs. (3.17), (3.21), and (3.22), the impacts of W_p on the booster capital cost, booster operation and maintenance cost, and energy cost are undeniable. So, a lower value of accumulative W_p may decrease the total cost.

In addition, increasing the value of maximum allowable diameter, $D_i^{\max}$, may conclude less expensive solutions. Figure 3.7 demonstrates the optimal results for a trunk pipeline with a maximum allowable diameter of 1 m and different number of booster stations.

Furthermore, Table 3.4 represents the optimal design parameters for $D_i^{\max} = 1$ m. As it can be observed in Fig. 3.7, the total costs for the first three different cases are

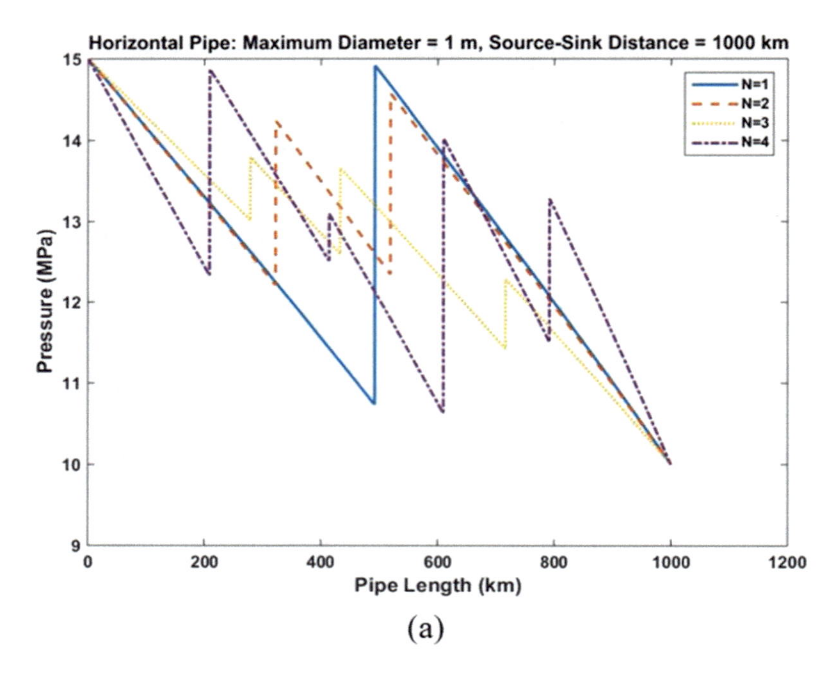

(a)

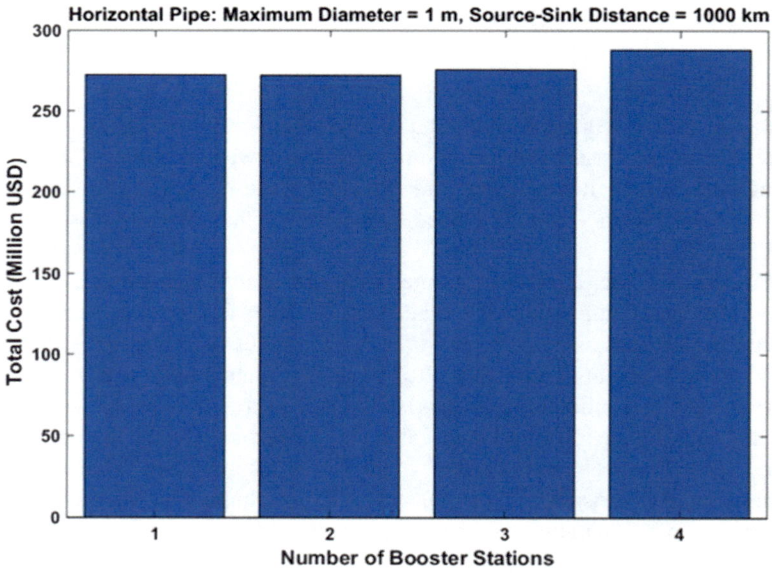

(b)

Fig. 3.7 (**a**) CO_2 pressure variation and (**b**) total cost for the optimal design of a 1000 km horizontal pipeline with different number of booster stations. $D_i^{max} = 1$ m, mass flowrate = 750 kg/s, $O\&M_{booster} = 3.25\%$ of $Cap_{booster}$

Table 3.4 Optimal design parameters for a 1000 km horizontal pipeline. $D_i^{max} = 1$ m, mass flowrate = 750 kg/s, $O\&M_{booster} = 3.25\%$ of $Cap_{booster}$

	Booster station power (MW)				Pipeline segment diameter (m)					Pipeline segment length (km)				
Number of boosters	Wp_1	Wp_2	Wp_3	Wp_4	D_1	D_2	D_3	D_4	D_5	L_1	L_2	L_3	L_4	L_5
1	5.2628	–	–	–	0.92	0.91	–	–	–	492	508	–	–	–
2	2.4636	2.7021	–	–	0.92	0.90	0.91	–	–	323	196	481	–	–
3	0.9334	1.2797	1.0680	–	0.95	0.94	0.94	0.94	–	279	154	283	284	–
4	3.0850	0.7094	4.2861	2.1978	0.85	0.87	0.86	0.84	0.83	209	205	196	182	208

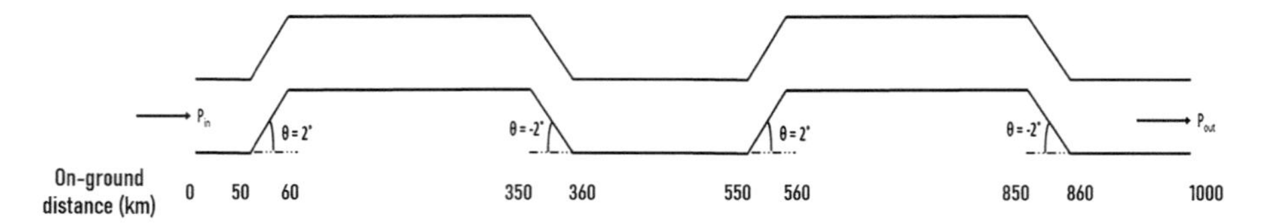

Fig. 3.8 Schematic of a pipeline with uphill and downhill slopes

almost equal. So, it is justifiable to choose a three-booster station scenario, based on reliability and availability concepts. Moreover, by comparing Figs. 3.6 and 3.7, one can conclude that increasing the maximum allowable pipeline diameter may reduce the total cost of the best solution. For instance, the total cost of the best solution corresponding to a maximum allowable diameter equalling 1 m (272.19 M$) is around 85% of that corresponding to a maximum allowable diameter of 0.8 m (321.60 M$).

Case Study 2: Trunk Pipeline with Uphill and Downhill Slopes
In reality, CO_2 transportation routes between source-sink pairs are seldom purely horizontal. They are full of elevation changes at several locations. This also increases the total pipe length. As any changes in pipeline slope have impacts on CO_2 pressure drop, corresponding modifications in system design parameters like booster stations' power and pipeline segments' diameter are necessary to guarantee the required pressure. As an example, to show the impacts of topography on the optimal solution, a case study with uphill and downhill slopes of 2° is considered (Fig. 3.8). In this pipeline, the CO_2 mass flowrate is 750 *kg/s* and the maximum allowable diameter equals 1 *m*.

The pressure and density profiles corresponding to the optimal solutions are depicted in Fig. 3.9. Clearly, a drop in pressure causes a fall in density. The CO_2 density remains between 750 and 870 kg/m^3 along the pipeline. The pressure profile shows any changes in pipeline slope can substantially influence the pressure drop: a 2° uphill slope over an on-ground distance of 10 km (sloping segment length = 10/cos 2° km) causes a 3 *MPa* pressure drop. On the other hand, a downhill slope of the same specifications boosts CO_2 pressure by around 2.5 *MPa*.

Table 3.5 represents the optimal parameters for this example. By comparing the locations of booster stations in the sloped pipeline and in the horizontal pipeline, it can be concluded that for the sloped pipeline, the booster stations are placed near the uphill slope(s) to compensate for the augmented pressure drop. Obviously, while determining optimal pipeline routes, downhill terrain slopes are the best options as they behave as booster stations, while uphill slopes must be avoided. Applying this information in designing the pipeline route will help reduce the total costs. Figure 3.10 depicts the total cost for different number of booster stations. The least cost is associated with the solution corresponding to two booster stations.

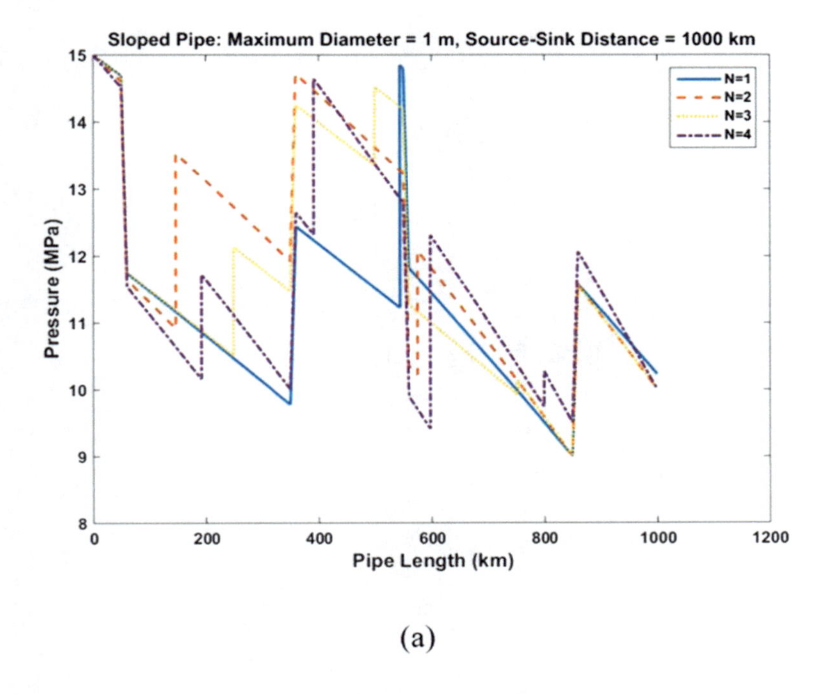

(a)

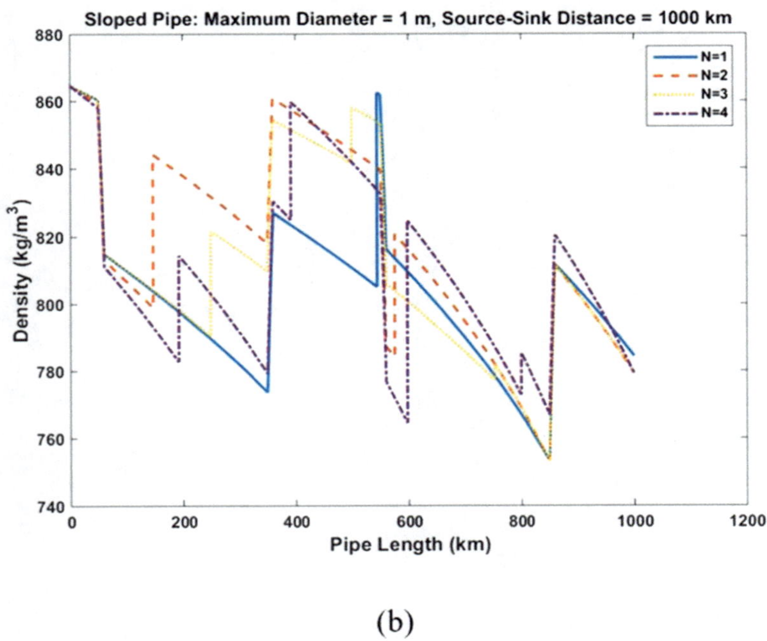

(b)

Fig. 3.9 (**a**) Pressure variation and (**b**) density variation of CO_2 for the optimal design of a sloped pipeline connecting a source and sink 1000 km apart with different number of booster stations. $D_i^{max} = 1$ m, mass flowrate = 750 kg/s, $O\&M_{booster} = 3.25\%$ of $Cap_{booster}$

Table 3.5 Optimal design parameters for a sloped pipeline connecting a source and sink 1000 km apart. $D_i^{max} = 1$ m, mass flowrate = 750 kg/s, $O\&M_{booster} = 3.25\%$ of $Cap_{booster}$

	Booster station power (MW)				Pipeline segment diameter (m)					Pipeline segment diameter (km)				
Number of boosters	Wp_1	Wp_2	Wp_3	Wp_4	D_1	D_2	D_3	D_4	D_5	L_1	L_2	L_3	L_4	L_5
1	4.5013	–	–	–	0.98	0.91	–	–	–	544	456	–	–	–
2	3.2860	2.3659	–	–	0.93	0.94	0.89	–	–	147	429	424	–	–
3	2.0561	1.3726	0.2710	–	0.98	0.98	0.97	0.89	–	249	250	252	248	–
4	2.0033	2.8188	3.7983	0.6954	0.90	0.89	0.87	0.86	0.84	192	199	206	201	200

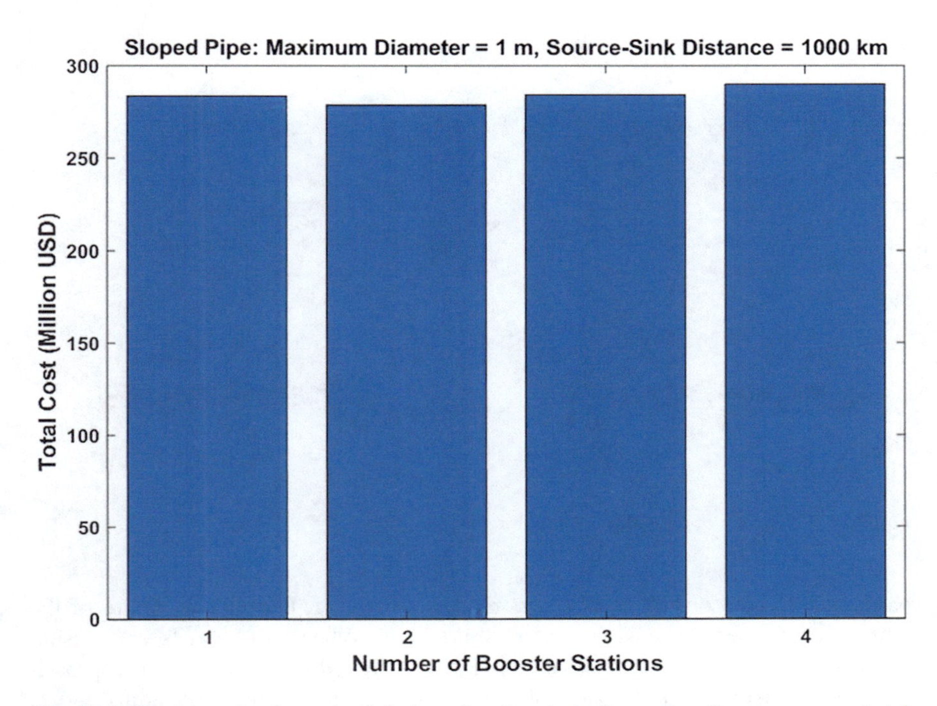

Fig. 3.10 Total cost for the optimal design of a sloped pipeline connecting a source and sink 1000 km apart with different number of booster stations. $D_i^{max} = 1$ m, mass flowrate = 750 kg/s, $O\&M_{booster} = 3.25\%$ of $Cap_{booster}$

3.4 Conclusions

In this study, a framework has been presented to minimize the cost of a CO_2 transportation system, while considering several design parameters and constraints (e.g., topographical settings, the pipeline's source and sink pressures, number of booster stations, and pipeline segment diameters). The obtained results show that the pipeline diameter and slope are the key design factors. Clearly, the optimization exercise, which determines the number and capacity of the booster stations and the diameter and length of each pipeline segment, depends on the design and operating limitations, and uphill and downhill slopes along the path. As an example, for pipes with sloping sections, booster stations are best located near the uphill slopes. Moreover, the CO_2 density and pressure profiles change along the pipeline, which makes it imperative to include in the model expressions used in their calculation. In addition, to decrease the overall cost of the transportation system, wherever possible, the optimal pipeline routes should use downhill slopes of the terrain to reduce the inherent need to add auxiliary booster stations. Although in this study we focused on pipelines carrying pure CO_2, in reality, impure CO_2 causing variations in density along the pipeline ought to be transported. Furthermore, we discussed examples with

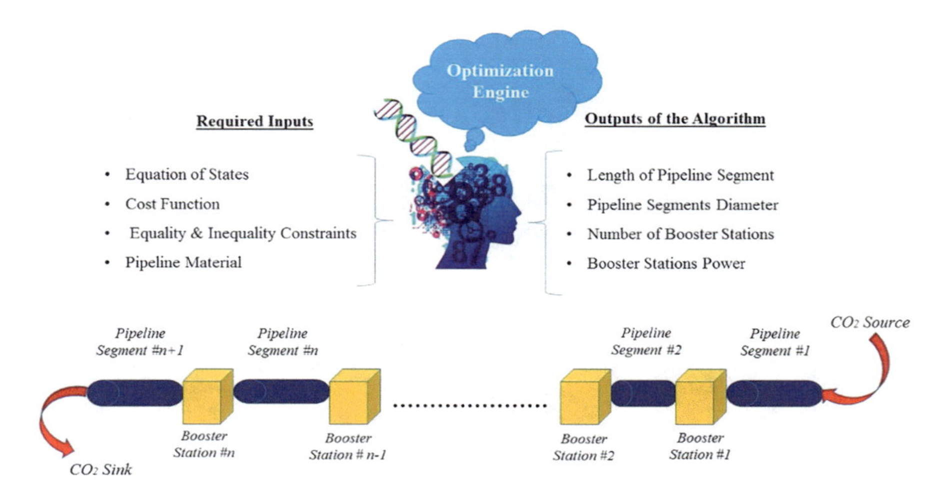

Fig. 3.11 Schematic of the proposed solution [48]

particular design constraints and cost models. However, the generic nature of the introduced framework ensures compatibility with other cost models and design constraints to seek optimal solutions. Finally, the utilized GA-based approach generally converges to almost global optimum results by properly adjusting its parameters. Consequently, the presented method is a powerful technique to solve the CO_2 transportation optimization problem. The graphical summary of the presented solution is shown in Fig. 3.11.

References

1. B.J. van Ruijven, D.P. van Vuuren, J. van Vliet, A.M. Beltran, S. Deetman, M.G.J. den Elzen, Implications of greenhouse gas emission mitigation scenarios for the main Asian regions. Energy Econ. **34**(Supplement 3), S459–S469 (2012)
2. IPCC, Climate change 2022: impacts, adaptation, and vulnerability, in *Contribution of Working Group II to the Sixth Assessment Report of the Intergovernmental Panel on Climate Change*, ed. by H.-O. Pörtner, D.C. Roberts, M. Tignor, E.S. Poloczanska, K. Mintenbeck, A. Alegría, M. Craig, S. Langsdorf, S. Löschke, V. Möller, A. Okem, B. Rama, (Cambridge University Press, Cambridge/New York, 2022), 3056 pp. https://doi.org/10.1017/9781009325844
3. B. Metz, O. Davidson, H. de Coninck, M. Loos, L. Meyer, *Carbon Dioxide Capture and Storage* (Intergovernmental Panel on Climate Change, 2015)
4. R. Svensson, M. Odenberger, F. Johnsson, L. Strömberg, Transportation systems for CO2—application to carbon capture and storage. Energy Convers. Manage. **45**(15–16), 2343–2353 (2004)
5. A. Witkowski, A. Rusin, M. Majkut, S. Rulik, K. Stolecka, Comprehensive analysis of pipeline transportation systems for CO2 sequestration. Thermodynamics and safety problems. Energy Convers. Manage. **76**, 665–673 (2013)

6. P.N. Seevam, M.J. Downie, J.M. Race, *Transport of CO2 for Carbon Capture and Storage in the UK* (Society of Petroleum Engineers)
7. K. Damen, M. van Troost, A. Faaij, W. Turkenburg, A comparison of electricity and hydrogen production systems with CO2 capture and storage—Part B: Chain analysis of promising CCS options. Prog. Energy Combust. Sci. **33**(6), 580–609 (2007)
8. R.S. Middleton, J.M. Bielicki, A scalable infrastructure model for carbon capture and storage: SimCCS. Energy Policy **37**(3), 1052–1060 (2009)
9. D. Zhao, Q. Tian, Z. Li, Q. Zhu, A new stepwise and piecewise optimization approach for CO2 pipeline. Int. J. Greenhouse Gas Control. **49**, 192–200 (2016)
10. Q. Tian, D. Zhao, Z. Li, Q. Zhu, Robust and stepwise optimization design for CO2 pipeline transportation. Int. J. Greenhouse Gas Control. **58**, 10–18 (2017)
11. P. Noothout, F. Wiersma, O. Hurtado, D. Macdonald, J. Kemper, K. van Alphen, CO2 Pipeline infrastructure – lessons learnt. Energy Procedia. **63**, 2481–2492 (2014)
12. W. Mallon, L. Buit, J. van Wingerden, H. Lemmens, N.H. Eldrup, Costs of CO2 transportation infrastructures. Energy Procedia. **37**, 2969–2980 (2013)
13. M.M.J. Knoope, A. Ramírez, A.P.C. Faaij, Economic optimization of CO2 pipeline configurations. Energy Procedia. **37**, 3105–3112 (2013)
14. G. Heddle, H. Herzog, K. Michael, *The Economics of CO2 Storage* (Massachusetts Institute of Technology, 2003)
15. L. Gao, M. Fang, H. Li, J. Hetland, Cost analysis of CO2 transportation: Case study in China. Energy Procedia. **4**, 5974–5981 (2011)
16. S.T. McCoy, E.S. Rubin, An engineering-economic model of pipeline transport of CO2 with application to carbon capture and storage. Int. J. Greenhouse Gas Control. **2**(2), 219–229 (2008)
17. J. Serpa, J. Morbee, E. Tzimas, Technical and economic characteristics of a CO2 transmission pipeline infrastructure, in *JRC Scientific and Technical Reports*, (European Comission, 2011)
18. J.J. Dooley, R.T. Dahowski, C.L. Davidson, S. Bachu, N. Gupta, J. Gale, A CO2 storage supply curve for North America and its implications for the deployment of carbon dioxide capture and storage systems, in *7th International Conference on Greenhouse Gas Control Technologies*, 2004. College Park
19. D.L. McCollum, J.M. Ogden, *Techno-Economic Models for Carbon Dioxide Compression, Transport, and Storage & Correlations for Estimating Carbon Dioxide Density and Viscosity* (Institute of Transportation Studies, University of California-Davis, 2006)
20. M.M.J. Knoope, A. Ramírez, A.P.C. Faaij, A state-of-the-art review of techno-economic models predicting the costs of CO2 pipeline transport. Int. J. Greenhouse Gas Control. **16**, 241–270 (2013)
21. M. van den Broek, A. Ramírez, H. Groenenberg, F. Neele, P. Viebahn, W. Turkenburg, A. Faaij, Feasibility of storing CO2 in the Utsira formation as part of a long term Dutch CCS strategy: an evaluation based on a GIS/MARKAL toolbox. Int. J. Greenhouse Gas Control. **4**(2), 351–366 (2010)
22. D.J. Zigrang, N.D. Sylvester, Explicit approximations to the solution of Colebrook's friction factor equation. AIChE J. **28**(3), 514–515 (1982)
23. T. Wildenborg, C. Hendriks, S. Holloway, R. Brandsma, E. Kreft, A. Lokhorst, Cost curves for CO2 storage: European sector, in *7th International Conference on Greenhouse Gas Control Technologies*, (2004)
24. Z. Dongjie, W. Zhe, S. Jining, Z. LiLi, L. Zheng, Economic evaluation of CO2 pipeline transport in China. Energy Convers. Manage. **55**, 127–135 (2012)
25. T. Kazmierczak, R. Brandsma, F. Neele, C. Hendriks, Algorithm to create a CCS low-cost pipeline network. Energy Procedia **1**(1), 1617–1623 (2009)
26. H.Y. Benson, J.M. Ogden, Mathematical programming techniques for designing minimum cost pipeline networks for CO2 sequestration, in *6th International Conference on Greenhouse Gas Control Technology*, (2003)
27. D.E. Goldberg, *Genetic Algorithms in Search, Optimization and Machine Learning* (Addison-Wesley Longman Publishing Co., Inc., 1989), p. 372

28. M. Mitchell, *An Introduction to Genetic Algorithms* (MIT Press, 1998)
29. T. Murata, H. Ishibuchi, H. Tanaka, Multi-objective genetic algorithm and its applications to flowshop scheduling. Comput Ind. Eng. **30**(4), 957–968 (1996)
30. B. Sahiner, H.P. Chan, N. Petrick, M.A. Helvie, M.M. Goodsitt, Design of a high-sensitivity classifier based on a genetic algorithm: application to computer-aided diagnosis. Phys. Med. Biol. **43**(10), 2853–2871 (1998)
31. S.J. Montoya, W.A. Jovel, J.A. Hernandez, C. Gonzalez, Genetic Algorithms Applied to the Optimum Design of Gas Transmission Networks. SPE International Petroleum Conference and Exhibition in Mexico, 1–3 February, Villahermosa, Mexico 2000. Society of Petroleum Engineers.
32. D.A. Savic, G.A. Walters, Genetic algorithms for least-cost design of water distribution networks. J. Water Res. Plan. Man. **123**(2), 67–77 (1997)
33. T.D. Prasad, N.S. Park, Multiobjective genetic algorithms for design of water distribution networks. J. Water Res. Plan. Man. **130**(1), 73–82 (2004)
34. B. Tolson, H.R. Maier, A.R. Simpson, B.J. Lence, Genetic algorithms for reliability-based optimization of water distribution systems. J. Water Res. Plan. Man. **130**(1), 63–72 (2003)
35. J. van Zyl, D. Savic, G. Walters, Operational optimization of water distribution systems using a hybrid genetic algorithm. J. Water Res. Plan. Man. **130**(2), 160–170 (2004)
36. Z.X. Zhang, G.X. Wang, P. Massarotto, V. Rudolph, Optimization of pipeline transport for CO2 sequestration. Energy Convers. Manage. **47**(6), 702–715 (2006)
37. M.K. Chandel, L.F. Pratson, E. Williams, Potential economies of scale in CO2 transport through use of a trunk pipeline. Energy Convers. Manage. **51**(12), 2825–2834 (2010)
38. P.N. Seevam, J.M. Race, M.J. Downie, P. Hopkins, Transporting the Next Generation of CO2 for Carbon, Capture and Storage: The Impact of Impurities on Supercritical CO2 Pipelines, in Proceedings of the ASME International Pipeline Conference, 2008.
39. B. Wetenhall, H. Aghajani, H. Chalmers, S.D. Benson, M.C. Ferrari, J. Li, J.M. Race, P. Singh, J. Daviso, Impact of CO2 impurity on CO2 compression, liquefaction and transportation. Energy Procedia **63**, 2764–2778 (2014)
40. H. Li, J. Yan, M. Anheden, Impurity impacts on the purification process in oxy-fuel combustion based CO2 capture and storage system. Appl. Energy. **86**(2), 202–213 (2009)
41. GlobalEnergyObservatory. Current List of Coal Power Plants. Available from: http://globalenergyobservatory.org/list.php?db=PowerPlants&type=Coal
42. PowerTechnology.com. 'Giga' projects – the world's biggest thermal power plants. Available from: http://www.power-technology.com/features/feature-giga-projects-the-worlds-biggest-thermal-power-plants
43. K. Patchigolla, J.E. Oakey, Design overview of high pressure dense phase CO2 pipeline transport in flow mode. Energy Procedia. **37**, 3123–3130 (2013)
44. M. Nazeri, A. Chapoy, R. Burgass, B. Tohidi, Measured densities and derived thermodynamic properties of CO2-rich mixtures in gas, liquid and supercritical phases from 273 K to 423 K and pressures up to 126 MPa. J. Chem. Thermodyn **111**, 157–172 (2017)
45. A. Fenghour, W.A. Wakeham, The viscosity of carbon dioxide. J. Phys. Chem. Ref. Data. **27**, 31 (1998)
46. A. Chipperfield, P.J. Fleming, H. Pohlheim C.M. Fonseca, Genetic Algorithm Toolbox for use with Matlab. Technical Report No. 512, 1994. Department of Automatic Control and Systems Engineering, University of Sheffield.
47. S.N. Sivanandam, S.N. Deepa, *Introduction to Genetic Algorithms* (Springer, Berlin/Heidelberg, 2008)
48. M. Mohammadi, F. Hourfar, A. Elkamel, Y. Leonenko, Economic optimization design of CO2 pipeline transportation with booster stations. Ind. Eng. Chem. Res. **58**(36), 16730–16742 (2019)

Chapter 4
Power-to-X and Electrification of Chemical Industry

Kelly Wen Yee Chung, Sara Dechant, Young Kim, Ali Ahmadian, and Ali Elkamel

4.1 Introduction

Increasing global energy consumption and a call to act against climate change have led to a search for alternative electricity generation. Implementing more renewable energy sources will have a positive impact on the chemical industry, reducing its reliance on greenhouse gas (GHG)-emitting energy sources. Power-to-X technologies provide energy storage options for the energy produced from these renewable sources. The purpose of this chapter is to review several sources related to Power-to-X and the electrification of the chemical industry and provide information on the current state of this research, any gaps in knowledge, and how these topics can progress further in the future.

4.1.1 Current Power-to-X Projects

In May 2020, a green hydrogen project under the name of HYFLEXPOWER was launched [1]. This project is a consortium made up of Engie Solutions, Siemens Gas

K. W. Y. Chung · S. Dechant · Y. Kim
Department of Chemical Engineering, University of Waterloo, Waterloo, ON, Canada

A. Ahmadian (✉)
Department of Electrical Engineering, University of Bonab, Bonab, Iran

Department of Chemical Engineering, University of Waterloo, Waterloo, ON, Canada

A. Elkamel
Department of Chemical Engineering, University of Waterloo, Waterloo, ON, Canada

Department of Chemical Engineering, Khalifa University, Abu Dhabi, United Arab Emirates
e-mail: aelkamel@uwaterloo.ca

A. Ahmadian et al. (eds.), *Carbon Capture, Utilization, and Storage Technologies*, Green Energy and Technology, https://doi.org/10.1007/978-3-031-46590-1_4

and Power, Arttic, German Aerospace Center (DLR), and four European universities. HYFLEXPOWER aims to construct the world's very first industrial-scale Power-to-X-to-Power demonstrator with an advanced hydrogen turbine (Siemens SGT-400 industrial gas turbine). This project aims to prove that hydrogen can be produced and stored from renewable electricity and then added with up to 100% to the natural gas currently used with combined heat and power plants. The installed demonstrator will be used to store excess renewable electricity in the form of green hydrogen which indicates the first part of Power-to-X. During periods of high demand, this stored green hydrogen will then be used to generate electrical energy to be fed into the grid (X-to-Power). The gas turbine will be used to convert stored hydrogen into electricity and thermal energy. This implementation is expected to reduce 65,000 tons of CO_2 per year [2].

Along with other industrial and academic partners, applied knowledge institute TNO has launched a research program called "Voltachem," to study innovative means on the use of sustainable electricity in the chemical industry [3]. The program addresses both the indirect and direct use of electricity within the chemical industry and works from a systemic point of view. The activities span across four program lines, which are power-2-integrate which consists of exclusive discussion and group work on roadmap and system studies, Power-to-Heat, power-2-hydrogen, and Power-to-Chemicals, which will be discussed further in this chapter.

As the Dutch industry accounts for approximately 46% of total energy use in the Netherlands, it plays an important role in reducing carbon emission and also in the energy transition toward sustainability. The Dutch industry also contributes to the climate goals made during the Paris Agreement through electrification of processes. An in-depth study report of promising transition pathways and innovation opportunities for electrification in the Dutch process industry was carried out by Berenschot, CE Delft, Industrial Energy Experts, and Energy Matters, in cooperation with TKI Industry and Energy [4]. According to the report, new and innovation systems and processes are vital. This includes improved high temperature heat pumps, new business models, and more room and support for experimentations.

A report has been created by Frontier Economics and published on behalf of the Weltenergieriat-Deutschland, along with other project partners around the world on the International Aspects of Power-to-X Roadmap [5]. The report focuses on the production of fuels generated from renewable electricity (Power-to-X) and determined the major pillars of a roadmap toward a global Power-to-X market. The three major pillars were:

- Enhance and support the scaling up of PtX technologies to achieve significant cost savings and promote international trade.
- Ensure reliable demand structures and spur the growth of the global PtX market.
- Facilitate an adequate framework for investments.

Table 4.1 Keywords

Keywords
Power-to-X
Sustainability
Green energy
Electrification
Chemical industry

4.2 Methodology

The methodology of this literature review chapter is based on numerous existing articles related to Power-to-X and the electrification of the chemical industry. A set of keywords were applied and used to complete this review chapter. Table 4.1 shows the keywords used to study this topic. Most of the articles were searched through Science Direct by using a combination of the keywords presented in the table. For more precise topics such as the electrolysis of water, other keywords are added to the base keywords we have identified. Other keywords such as ammonia processing, compressed-air energy systems (CAES), and electrocatalysts were necessary to guide us to the appropriate articles.

To ensure that the papers related to our research topic, approximately 100 articles were screened. Other articles were also found through ResearchGate. Connections between different articles were determined not only from the search engine but also through the references cited in the papers. Some topics needed more in-depth readings and comprehension; therefore, articles were directed first to understanding the state-of-art before furthering the research such as the current state of research, techno-economic assessments, and numerical and computational approach.

A total of 45 were used to study the Power-to-X and Electrification of Chemistry topic. In the literature classification section, we have grouped a list of references mainly based on the studies of the electrification technologies according to the region, state, research topic, publication year, scale (industrial or lab-scale), and the highlights extracted from the different articles. Most of the articles are dated after 2015, to provide a more modern insight into the current state of research and future developments of the technologies. As the topic is considered relatively new, more than half of the articles presented here are laboratory-scale.

4.3 Power-to-X and Electrification of the Chemical Industry

Electrification is a promising solution that offers the possibility of reducing the carbon emission of chemical industries, especially in sectors that depend heavily on fossil fuel sources. Power-to-X technologies offer great strategies for the massive storage of electricity generated from renewable sources. Power-to-X technology refers to the transformation and conversion of electricity. It is (generally of renewable origin) in another energy carrier or its storage. This new vector can be "heat" to meet industrial needs or heat supplies. It can also be a gas such as hydrogen or methane.

There are various ways for specific categories of Power-to-X technologies. This report will classify PtX technologies in 3 categories of:

1. Power-to-Heat
2. Power-to-Gas
3. Power-to-Chemicals

4.3.1 Power-to-Heat

Power-to-Heat is expected to be the first type of electrification that will be implemented on a large scale. Electricity is either used directly to generate heat or to upgrade heat and steam for more efficient use in chemical processes. Power-to-Heat can be applied in drying operations, process heat operations, such as steam and hot water, thermal, melting and casting, and also sterilization and pasteurization unit operations. A simple example would be the electrically driven heat pump, which can offer significant carbon reduction compared to fossil fuel combustion. Power-to-Heat can also be applied in induction furnaces, microwave heating, electric plasma heating, infrared heating, impulse drying, etc. [6].

When renewable energy supply surpasses demand, Power-to-Heat systems can use the surplus of energy to supply industrial needs, therefore avoiding the curtailment of renewable energy. Other advantages include increased flexibility through load shifting and energy storage on large scales.

Power-to-Heat analytical models have proven effective for residential heating applications [7]. Heat pumps and passive thermal storage are two Power-to-Heat technologies that efficiently transform energy into heat. Figure 4.1 shows the various means of converting electricity into heat in residential applications.

In centralized Power-to-Heat, electricity is converted to heat at a location other than where the heat will be applied. District heating then distributes the heat to where it is needed. Decentralized Power-to-Heat converts electricity to heat at the location where the heat is being applied. Centralized heating options are required to have thermal energy storage capabilities, whereas decentralized options are not, due to their proximity to the location of heat application. This is called direct heating. If a decentralized option is combined with thermal energy storage (TES), it is called TES-coupled heating. This system can be internal or external depending on where the Power-to-Heat element is located. External storage can be further broken down into heat pumps and resistive heaters. Figure 4.2 depicts the connection between Power-to-Heat options, energy sources, and heating networks.

A study on the impact of Power-to-Heat for the Dutch chemical industry showed that Power-to-Heat technologies for producing heat up to 200 °C could result in about 15–20% energy savings for heat demand, which leads to a reduction of carbon

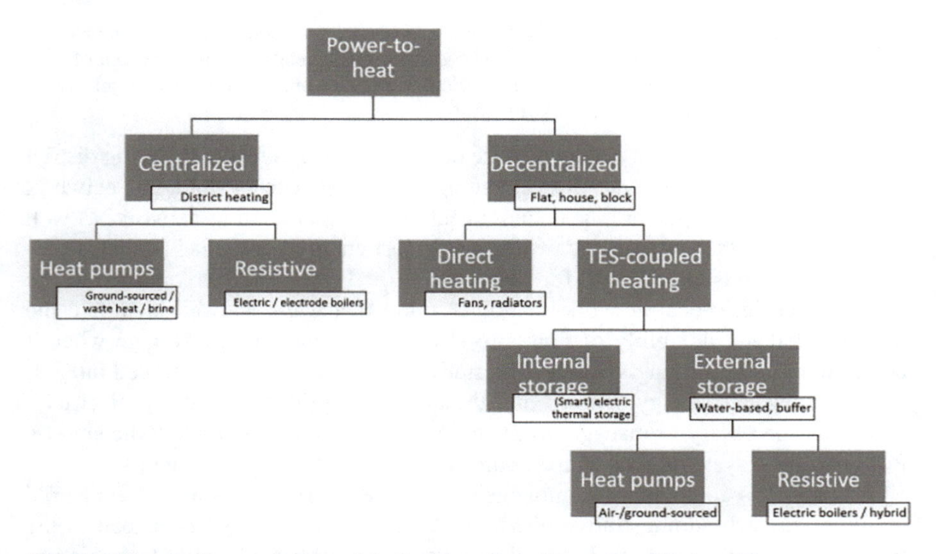

Fig. 4.1 Means of converting electricity into heat for residential applications [7]

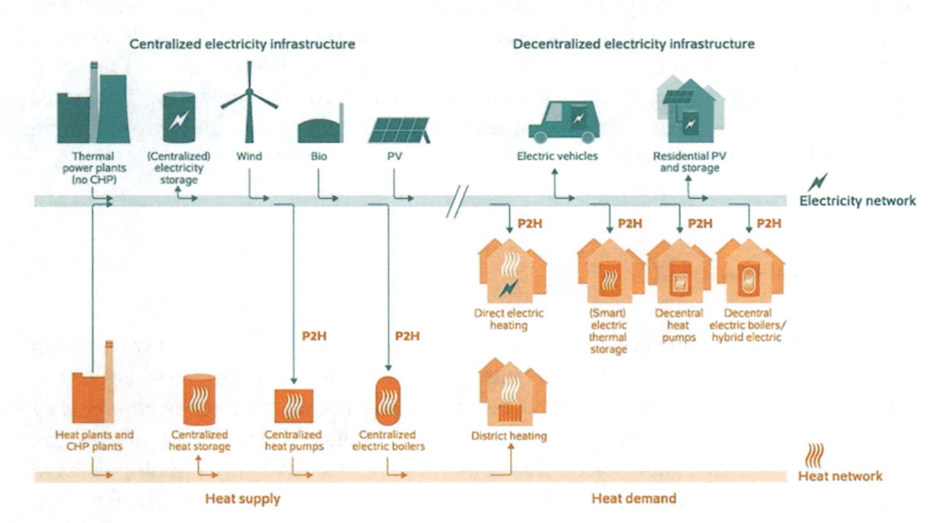

Fig. 4.2 Connection between Power-to-Heat options, energy sources, and heating networks

dioxide emission of about 6Mtonnes [4]. It could also offer a more flexible load shifting to help compensate for fluctuations of renewable sources. A study was completed on the coupling of local district heating and electrical distribution grids, which presents a method for a technical assessment on such a coupling. The study stated:

> This work presents a method that enables a detailed technical assessment of the operation of such coupled heat and power networks. It is based on a sequential coupling approach of a dynamic thermal-hydraulic model for the district heating network and a quasi-static model for the electrical distribution network [8]

The study showed success in coupling the networks and, as a result, lowering district heating supply temperatures, implementing renewable energy into the network, introducing low-temperature heat sources into the district heating network, as well as other effects that support a transition to "smart energy networks" [8]. This provides an optimistic outlook for the future of Power-to-Heat options to be implemented into local networks. Another study that was reviewed identified the need for demand flexibility of Power-to-Heat and thermal energy storage when it comes to building heating systems. The study states that with the continued integration of renewable energy sources into the electricity grid, flexibility in electricity demand is necessary to manage power grids. This flexibility is largely dictated by Power-to-Heat systems such as heat pumps and thermal energy storage [8].

A study was completed on implementing Power-to-Heat systems in residential buildings in each administrative district in Germany. The overall heat load in the buildings and how much of that load is covered by electrical heating technologies were quantified. The analysis showed that there is a greater amount of small and large central heating in cities, whereas there is a greater amount of medium-scale central heating in rural areas. The study then investigated heat pumps and resistive heaters in residential buildings. This data is essential for implementing Power-to-Heat systems. If Germany were to implement these systems, this study would prove effective for future calculations.

4.3.2 Power-to-Gas

The relation between electric power and natural gas has always been one way, which is gas-to-power (GtP). In recent years, the inversed relation, thus power-to-gas (PtG), has been paid more attention due to the increasing numbers of renewable energy installations and its high storage capacity. This technology is studied to produce the well-known synthetic fuels, electrofuels, or efuels. Power-to-Gas technologies have a strong potential in reducing carbon emissions CO_2. According to Thema et al. (2019) [10], there are 153 ongoing Power-to-Gas demonstration plants. In Europe, most of the plants are located in Germany, Denmark, and the United Kingdom.

The Power-to-Gas application could happen in two ways [11]:

1. Through chemical approach, which consists of producing gas-to-electrochemical reactions or processes.
2. Via the potential approach: application of power to pressurize or change a phase of a gas, for an instance air or natural gas.

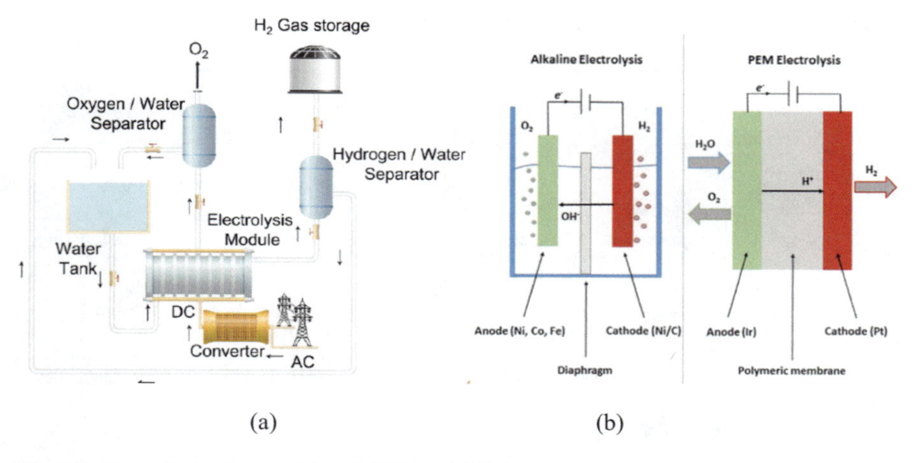

Fig. 4.3 Electrolysis of water (**a**) and AWE and PEM systems (**b**)

4.3.2.1 P2G Applications Via Chemical Approach

4.3.2.1.1 Power-to-Hydrogen

Power-to-Hydrogen falls under the category of Power-to-Gas systems, which consists of the production of gaseous fuel through electricity. Through an electrochemical reaction, namely, electrolysis of water (Fig. 4.3a), hydrogen can be produced by separating the hydrogen from oxygen in the water with an electricity input (Eq. 4.1). In Power-to-Hydrogen technology, the electrolyzer module is connected with the electricity grid. Because connecting to the electricity grid induces greenhouse gas emissions, a greener method such as connecting the electrolyzer to the wind turbine was studied and given more attention [12]:

$$\text{Anode} : \mathrm{H_2O} \rightarrow 1/2\mathrm{O_2} + 2\mathrm{H} + +2\mathrm{e}–$$

$$\text{Cathode} : 2\mathrm{H} + +2\mathrm{e}– \rightarrow \mathrm{H_2}$$

$$\text{Overall} : \mathrm{H_2O} \rightarrow \mathrm{H_2} + \frac{1}{2}\mathrm{O_2} \tag{4.1}$$

Different electrolytes systems were developed, such as alkaline water electrolysis (AWE), proton exchange membranes (PEM), and solid oxide water electrolysis (SOE). These systems operate at different temperature conditions. For an instance, solid oxide water electrolysis (SOE) operates at a higher range of temperatures (500 °C–1000 °C), whereas proton exchange membranes (PEMs) and alkaline anion exchange membranes (AEMs) operate at a temperature between 20 °C and 200 °C. Most of these electrolyte systems are built for laboratory scale, but many types of research have been studied to upgrade the electrolyte systems to the industrial scale.

Table 4.2 Techno-economic evaluation of AWE and PEM method

Technology		Alkaline		PEM	
	Unit	2017	2025	2017	2025
Efficiency	kWh of electricity/kg of H_2	51	49	58	52
Efficiency (LHV)	%	65	68	57	64
Lifetime Stack	Operating hours	80 000 h	90 000 h	40 000 h	50 000 h
CAPEX – total system cost (incl. power supply and installation costs)	EUR/kW	750	480	1200	700
OPEX	% of initial CAPEX/year	2%	2%	2%	2%
CAPEX – stack replacement	EUR/kW	340	215	420	210
Typical output pressure[a]	Bar	Atmospheric	15	30	60
System lifetime	Years	20		20	

Notes: H_2 = hydrogen; h = hour; kg = kilogram; kW = kilowatt; kWh = kilowatt hour; LHV = lower heating value; OPEX = operating expenditure; CAPEX and OPEX are based on a 20 MW system.

[a] Higher output pressure leads to lower downstream cost to pressurize the hydrogen for end use.

Alkaline water electrolysis is an advanced technology and is the first electrolysis system. It has been applied in the industry for over a century and is so far the cheapest way to produce hydrogen of electrolytic grade (~700-800€/kW). It consists of a cathode, an anode, and a thin porous ceramic diaphragm submerged in an alkaline electrolyte. The PEM method is also viewed as a promising technology toward upscaled hydrogen production. Unlike AWE, it does not use a liquid electrolyte but a thin solid polymer membrane.

In comparison to the technology readiness, alkaline water electrolysis (AWE) and power exchange membranes (PEM) are more technology-ready than other electrolyte systems on an industrial scale. S.A. Grigoriev et al. [13] made a techno-economic evaluation of these two promising technologies shown in Table 4.2.

The hydrogen can also be generated through steam-methane reforming (SMR) and subsequent water-gas shift reaction (WGS). The sequence of two methods which is currently dominating the hydrogen synthesis technique widely applied in chemical industries, however, relies heavily on fossil fuels. Therefore, the investigation of alternative hydrogen production method, namely, electrolysis of water, etc., is of the utmost importance among various research in P2X.

In comparison to the technology readiness, alkaline water electrolysis (AWE) and power exchange membranes (PEM) are more technology-ready than other electrolytes systems on an industrial scale. S.A. Grigoriev et al. [13] made a techno-economic evaluation of these two promising technologies shown in Table 4.2.

$$CH_4 + 3H_2O \longrightarrow CO + 3H_2 \tag{4.2}$$

$$CO + H_2O(g) \longrightarrow CO_2 + H_2 \tag{4.3}$$

The reaction mechanism of SMR (Eq. 4.2) and WGS (Eq. 4.3) show that methane, commonly derived from natural gas, is converted into hydrogen and carbon dioxide. The Haber Bosch (HB) process uses natural gas (methane) in an environment fueled by natural gas (high temperature and pressure) and produces a significant amount of CO_2. Many studies incorporate traditional SMR-WGS hydrogen synthesis process as an intermediate to the future hydrogen production technology by utilizing methane produced by Power-to-Methane as a feedstock to the reactions.

4.3.2.1.2 Other Applications of Hydrogen

The production of hydrogen can also be converted into methane which is used in the natural gas grid and also raw products such as ammonia and fuels through the addition of feedstocks such as carbon dioxide and nitrogen. After a post-treatment of the raw product, chemicals and fuel supply chains can be generated.

The formation of hydrogen H_2 through water electrolysis is the first part of the Power-to-Methane process chain. In the second step, CH_4 is formed via the reaction of H_2 with CO_2. The methanation reaction is shown in Eq. 4.4. CO_2 can be obtained from various sources, such as biomass plants, power generation plants, industrial processes, and ambient air.

$$4H_2 + CO_2 \longrightarrow CH_4 + 2H_2O \tag{4.4}$$

While electrolysis, at least for alkaline technology, is already mature and commercially available, the Power-to-Methane system still represents a relatively novel technology, aside from single commercial installation, which represents a technology readiness level (TRL) of about 5–7 [14].

Recently, Shell and ITM power plan to install a 10 MW electrolyzer at the Wesseling refinery site [15]. Proton Onsite and Siemens also recently delivered a MegaWatts scale PEM electrolyzer [16]. Thyssenkrupp has also recently come up with a large-scale hydrogen production that involves advanced water electrolysis using the "Zero-Gap" electrolysis technology. High efficiencies of more than 82% were achieved [17].

4.3.2.1.3 Studies on the PtG via Chemical Approach

One of the areas of research includes the conceptual framework to understand the P2G applications. Currently, the alkaline water electrolysis and proton exchange membrane electrolysis are industrialized for hydrogen production. Guo, Yujing, et al. [18] compared the working principles, advantages, and disadvantages of hydrogen production technologies. During his year of research (2019), alkaline water electrolysis with a flow rate of 1000 m3/h has been achieved whereas, in the case of proton exchange membrane technology, the most advanced equipment could produce 400 m3/h of hydrogen. However, alkaline water electrolysis showed downsides such as complex maintenance due to corrosion and slow start-up. Meanwhile, proton exchange membrane technology requires a much higher expenditure. A. Mazza et al. [19] did a comparison between the two technologies and on top of that a high-temperature steam electrolysis. Results showed higher efficiency of water decomposition at a higher temperature; therefore, this method has a high potential in the future. This technology is optimized for centralized and large-scale hydrogen production due to the fact that it requires a base-load operation and high-temperature steam. Bareib et al. [20] did a comparison on PEM water electrolysis and steam methane reforming through life cycle assessment. Results suggested that the PEM method is capable of reducing CO_2 emission by 75% and of global warming potential to 3.3 kg CO_2.

Currently, none of the literature describes the modeling of the whole electrolysis plant. Nevertheless, most of the modeling is based on the electrochemical behavior of the cell or stack. The modeling includes a mathematical description of the polarization curve using Faraday efficiency. An analytical model called the multi-physics models was developed by Hammoudi et al. [21] by taking into account all variations of structural parameters such as geometry parameters and operating conditions, unlike most conventional approaches which only takes into account the thermal aspect. This approach can be applied in a wide range of alkaline electrolyzers and one of the highlights of this approach is that the characterization can be completed in a relatively short time. The implementation of this model is made via MATLAB Simulink and a simulation tool validated by two industrial electrolyzers, which are Phoebus and Stuart. The model proposed obtained an accuracy with 0.9% relative deviation. One of the main findings in this article is the impact of operating pressure. A higher operating pressure decreases the power consumed by the electrolyzer and therefore the rate of hydrogen production.

Similar to Hammoudi et al.'s study, Henao et al. [22] also demonstrated a physics model for alkaline electrolyzers. The operating conditions (temperature, pressure, electrolyte concentration, and current) are used as an input for the model. A simulation tool called the alkaline electrolyzer simulation tool (AEST) incorporating the physics model and the electrical analogy was capable of describing the electrodes' physical and electrical performance and estimate the optimal operating current. AEST is a useful tool for predicting alkaline electrolyzer's performance and hydrogen production in long-term scenarios. Olivier et al. [23] developed a model under Bond Graph formalism which allows an accurate description of the dynamic behavior of a semi-industrial PEM electrolysis (25 kW). Currently, only a few models take into account all the energies involved in the electrolyte systems and its non-stationarity. The model is able to perform conception sensibility analysis and also develop a model-based diagnostic in the case of the presence of faults.

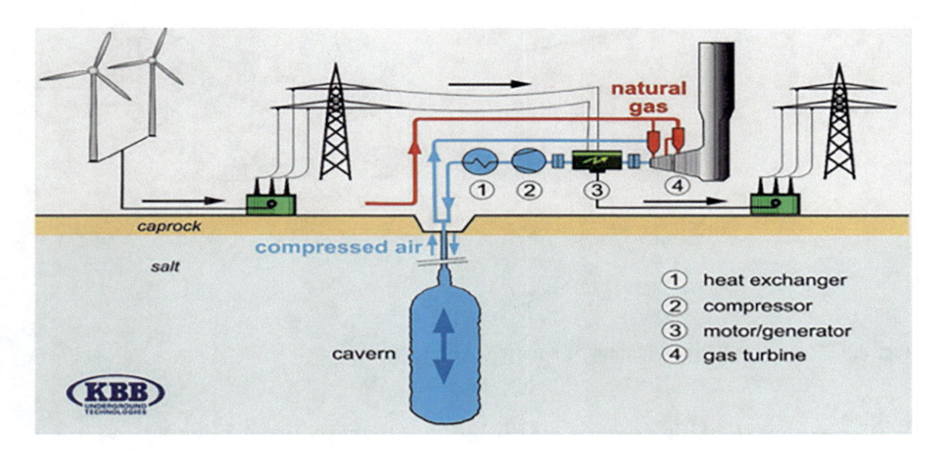

Fig. 4.4 An example of application of CAES system in a power plant

4.3.2.2 P2G Applications via Potential Approach

Unlike the chemical approach, this does not involve any chemical reaction. To compress air or change the phase of gas, one effective and promising electrical energy storage method is the Compressed-Air Energy Storage (CAES). The first plant of 290 MW (upgraded to 321 W in 2006) was built in Huntorf, Germany, in 1978. In 1991, another CAES plant of 100 MW was launched in McIntosh, USA. Although the Huntorf and McIntosh CAES power plants demonstrated maturity, the current development focuses on compressing processing without the need for fossil fuel burning [24, 25]. Figure 4.4 shows an example of the application of the CAES system in a power plant.

4.3.2.2.1 Compressed-Air Energy Storage (CAES)

CAES involves compression of air to store exergy and expanding the air to release exergy. The system has the capability to store zero net energy in the form of pressurized air. It is an attractive means for energy storage due to its constant pressure characteristic and also its merits of low cost and long design life.

Figure 4.5 shows a conceptual representation of a basic CAES system. Air flows through compressors driven by an electric motor before being stored in a cavern. The green thermal boxes could be simple coolers, heat exchangers, and straight ducts depending on the configuration of the CAES system. The preponderantly adiabatic process causes the pressure and temperature to increase during the passage through each compressor. When the grid needs power, the air is released from thus expanding through the expander and therefore impelling the electric generator and passing power back to the grid.

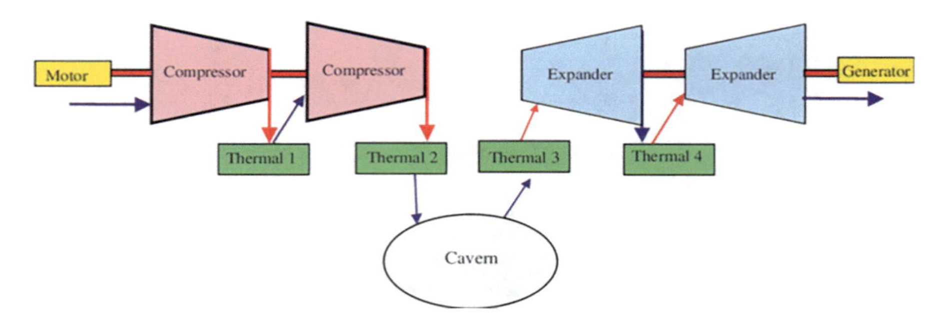

Fig. 4.5 Conceptual representation of a basic CAES system

The exergy stored in compressed air [26] is shown by Eq. 4.5 below:

$$B_{\mathrm{HP-air}} = (P_0 \times V_{\mathrm{store}}) \times \mathrm{rlogr}(r - 1) \quad (4.5)$$

where

r represents the pressure ratio,
P_0 represents the ambient pressure.
V_{store} represents the volume of high pressure stored.

For example, 41.m^3 of storage at $r = 200$ has the capability to store up to 1 MWh.

Although the two grid-scale plants (Huntorf and McIntosh) have been in operation for more than 20 years, they were not designed for the energy storage requirements of 2020 and 2030. The two grid-scale plants are diabatic in which fuel is used to heat the air instead of the expansion of the stored air itself; therefore, they are relatively inefficient. CAES still requires further development and research and therefore various forms of CAES systems were studied in recent years [24].

These included the adiabatic CAES (A-CAES) with thermal energy storage (TES). This type of energy storage does not require fuel supply and it has been reported as having high efficiency. A high-temperature A-CAES project named ADELE (TES with a temperature of ~600 °C) has an efficiency that reaches 70%. US-based LightSail Energy Ltd. also developed a CAES technology which came close to achieving isothermal conditions (I-CAES) [23].

4.3.2.2.2 Studies on the P2G via Potential Approach

To improve the efficiency of the CAES systems, Zhou et al. [27] conducted a comparative analysis of the different CAES systems. The performances of each system were compared using the round-trip efficiency (RTE) as an indicator, which evaluates the waste heat recovery and compression heat utilization. A thermodynamic model was developed and compared to the Huntorf plant. The study showed that the model proposed is valid as results showed less than 4% of the difference in the efficiency difference. A recuperator is added to the CAES system to recover waste heat. Results show that the exhaust temperature of the model has decreased from 482.2 °C to 59.4 °C and the efficiency is similar to the McIntosh plant. The study has shown that advanced A-CAES has a better efficiency compared to the conventional CAES which has a larger generating capacity at the cost of much waste heat and lower system efficiency.

Chen et al. [28] proposed an improved version of the conventional CAES system by applying a pre-cooling system to decrease energy consumption during the fourth compression stage. The thermodynamic model used in this study presents a similar concept to Zhou et al.'s model. The RTE was proved to increase by 3% compared to the conventional CAES systems. Five refrigerants (isobutane, R134a, R32, R1234yf, and R125) were studied for the application of the pre-cooling system. Isobutane presented the best choice among the five for refrigerant performance.

Moyazeni et al. [29] researched the effect of operating parameters such as the expansion flow rate on the performance of the A-CAES system. This system allows compression heat to be extracted during intercooling stages, and the heat is stored in a thermal energy reservoir to be reused to pressurize the air in the expanders. A dynamic model is used in this study. The study showed that a higher compressor flow rate reduces the system efficiency and increasing the expander flow rate reduces the system and heat recovery efficiencies up to 8.8–40.3%, thus showing the importance of compressor and expander flow selection.

4.4 Power-to-Chemical

4.4.1 Power-to-Specialties

The objective of Power-to-Specialties allows the production of chemical intermediates and higher value products via direct synthesis using electricity. Electrocatalytic conversion is required in the synthesis of chemical products with the addition of feedstocks such as biomass-derived products. The application of electrochemistry in this process has the major advantages of higher purity and selectivity compared to the usage of traditional processes. Other advantages include the possibility of ambient process conditions and a reduction in feedstock need, waste, and by-products.

4.4.2 Power-to-Commodities

Power-to-Commodities concerns both the centralized and decentralized production of large-volume chemicals. The profit margins are much lower than fine and specialty chemicals; thus, drivers of these processes are different than with Power-to-Specialties. Increased sustainability is the major reason for the current investigations into Power-to-Commodities processes.

4.4.3 Electrification of GHG-Intensive Chemical Processes

The optimization and imminent adaptation of newly developed technologies have led to 58% decrease in total GHG emissions in the EU chemical industry in the past 3 decades (Fig. 4.6) [30].

Further reduction of GHG in the chemical industry will require scientific breakthroughs inside the chemical processes that are heavily contributing to the majority of the remaining emissions. The top 5 chemical commodities that consume the most amount of energy and release GHG are ammonia, ethylene, propylene, methanol, and benzene/toluene/xylene (BTX) [31] Among these, ammonia production is one of the most energy-intensive chemical processes among various chemical processes that are currently practiced on an industrial scale. The Haber-Bosch (HB) process, the fundamental basis of ammonia production, converts nitrogen and hydrogen into ammonia that requires high pressure and temperature to yield optimal production of ammonia. The versatility of ammonia in various industries, most notably in fertilizer, led ammonia to be a high-volume commodity chemical represented by production volume exceeding 180Mt per year [32]. The huge carbon footprint associated with the consumption of natural gas in its process accounts for approximately 1.4% of global CO-2 emission. Its irreplaceable status in human civilization, amplified by its potential role in the upcoming hydrogen-based economy, inspired researchers around the world to investigate to reduce the GHG emission from the ammonia

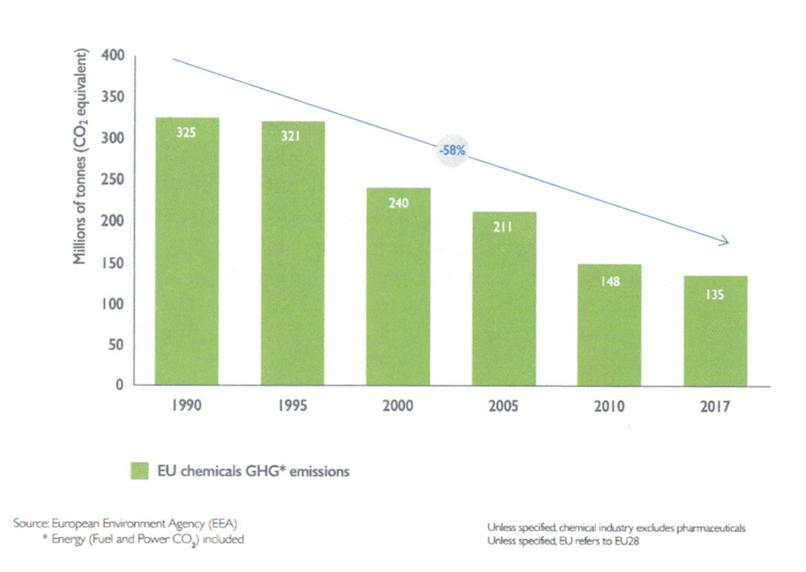

Fig. 4.6 Total GHG emissions in the EU chemical industry (1990 ~ 2017)

production plant. Among the various proposed approaches to the goal of reducing the carbon intensity of ammonia production, electrification via electrochemical synthesis of ammonia is surfacing as a potential alternative to replace the conventional HB process that has become essential in current society.

4.4.4 Empirical Development of Electrocatalysts

Still in the early stage of research, the current challenge in the transition to an alternative is identifying the electrocatalyst that meets the minimum requirement for the process to be applicable. Wu et al. provide a summary of the recent progress of catalysts studied for nitrogen reduction reactions. The literature concisely discusses the new approach and substances that were studied and evaluates the performance based on ammonia yield rate and Faradaic Efficiency (FE). Mainly, recent progress on the performance of electrocatalysts categorized as noble metals, non-noble metals, and non-metal elements is emphasized. The current research trend of noble metal (Au, Ru, Rh, Pd, Pt, Ag, and Ir) focuses on single atoms, nanostructures, and alloys due to the high cost of these elements. Recent notable success has been observed from hydrogen doping as the performance of Xu et al.'s nanoporous palladium has shown improvement in ammonia yield and mixed results in FE [33]. On the other hand, studies on non-noble metal (Y, Sc, Ti, Zr, V, Cr, Nb, Mo, Fe, Co, Mn, Ni, Cu, W, Re, Sn, Sb, Bi, La, Ce, and Dy) and non-metal (B, C, N, O, F, P, S, Se, and Te) elements are quantitatively more active due to much affordable cost. As for non-noble metal catalysts, Mo, Fe, and Cu-based catalysts are at the center of attention with many articles reporting performance improvement. As for non-metal catalysts, a wide range of investigations on F-doped carbon catalysts and black/red phosphorous are currently performed. Figure 4.7 shows the performance summary of the recently published study results with respect to ammonia yield and FE [34].

While the majority of the results exhibited improvement in either yield, FE, or both, all of the catalysts failed to produce the feasible industrial installation target imposed by the US Department of Energy REFUEL program ($9.3 * x10^{-7} \frac{\text{mol}}{\text{cm}^2 * \text{s}} \sim 57 \frac{\text{mg}}{\text{cm}^2 * \text{h}}, \text{FE} > 90\%$) [32]. The lack of systematic control over the experiment and unfavorable selectivity of the targeted reaction is considered to be the current obstacle at hand.

The study of electrocatalysts is not limited to ammonia. Various research is actively performed on developing electrocatalysts applicable in a different type of process like CO_2 reduction. Jhong et al. discuss the-state-of-the-art of the CO_2 reduction research on the progress of catalyst studies of the electrochemical process that converts CO_2 into multi-carbon fuels and chemicals. The study is designed to add value to the CO_2 that is obtained through the carbon-capturing technique which is currently removed from the carbon cycle by depositing in an underground geological formation. The successful implementation of CO_2 reduction into the

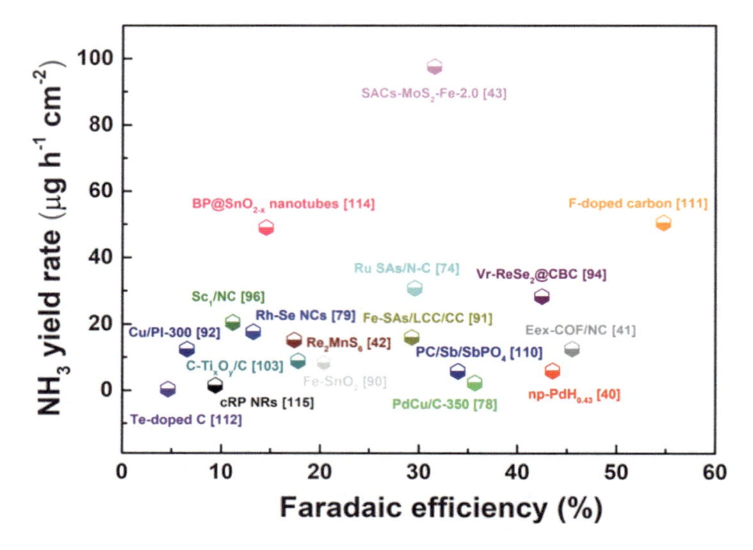

Fig. 4.7 Ammonia yield and Faradaic Efficiency of electro catalysts utilized in electrochemical ammonia synthesis

carbon capture technology will reduce the risks associated with the long-term deposition of carbon and further reduce the carbon emission as the product, synthetic fuel, can be utilized as an input to the plant's feedstock [35]. The current obstacle for the CO_2 reduction research comes from the wide range of reaction intermediates that are present between CO_2 and the targeted product. (Fig. 4.6) The selectivity issue of a high-value multi-carbon product becomes even more complex because the similarities in surface chemistry of various intermediate lead to a synergetic effect that increases the binding energy [36]. Approaches that are taken include the usage of bimetallic alloy catalysts and relation to its effect in breaking scaling relationship, manipulation of catalysts' oxidation state, and modification of catalysts surface to improve the performance of electrocatalysts used in electrochemical CO_2 reduction [37] (Fig. 4.8).

Another area of research is the electrolyte solution of the cells. In the "Study on POM assisted electrolysis for hydrogen and ammonia production," Zhao et al. expand added Silicotungstic acid (SiW12, redox medium) and organic solvents (regulates redox property of SiW12) to the solution used for electrolysis. Compared to pure water, the electrolyte solution showed a 35.31% increase in hydrogen yield and improvement in the ammonia yield and FE. The test also concludes that the presence of the ionic element commonly found in seawater did not affect the electrolysis [38].

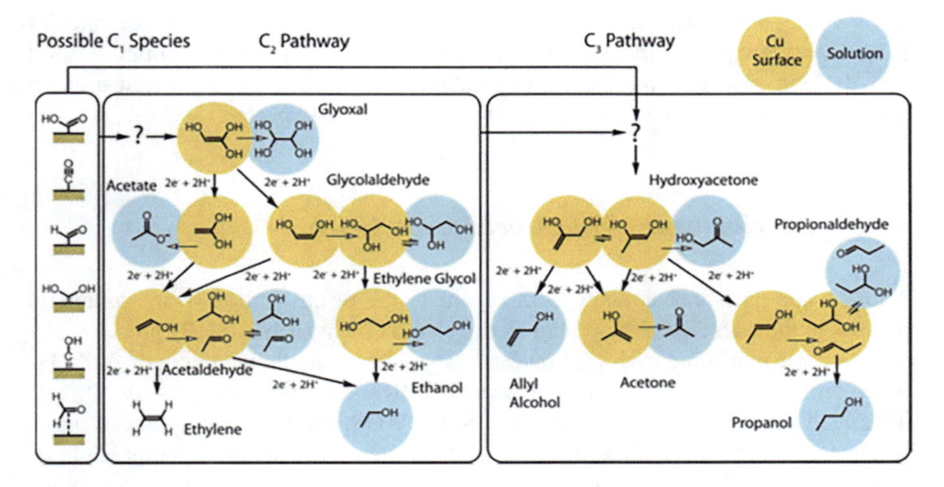

Fig. 4.8 Reaction pathway of carbon reduction reactions using copper catalyst

4.4.5 *Computational Studies on Electrocatalysts*

The advancement in computational modeling and hardware capacity allowed computational studies on the electrochemical reduction of nitrogen to be capable of performing complex calculations associated with the reaction. Catalyst research in the past decade has fully utilized quantum computational modeling based on density functional theory (DFT) to solve the complex problems in kinetic analysis, free-energy calculation, and catalyst design. The capability to comprehend the microscopic activities of catalytic reactions and interaction between the catalyst surface has been crucial in modern research [39]. A more recent trend applies to machine learning (ML) in studies that have proven to be highly suitable for the computational development of catalysis. For instance, Lu et al. in their studies on high-entropy alloys (HEA) that are gaining attention due to the high quantity of active-site used DFT to train the ML model. The model result concurred with the results from the experimental literature to a high degree of accuracy. Additionally, the model's high generalizability for bimetallic catalysts and simplicity suggests significance in identifying traits that are important in HEA catalyst design and expansion to research in other catalysts in the future research [40]. Kim et al. reported success in utilizing ML in evaluating the performance of the catalysts in the given database and suggested a set of catalysts that may exhibit better catalytic performance than the reference catalyst [41]. Zafari et al. utilized a deep neural network for screening the throughput yield of ammonia of the studied catalysts [42]. Wang et al. developed a new ML method applicable specifically to catalyst design. They used a dataset of 315 C1/C2 surface intermediates and transition states which all exert presence in a reaction pathway included in ethanol synthesis from synthetic gas. Three recently proposed methods of graph-convolution, weave, and graph neural networks were all selected for training the model that is built to predict the energies of surface species [43]. Considering that the limitation of the conventional computational method was its high-cost and time-consuming calculation process, further application of ML on the computational catalyst development is to be expected.

4.4.6 Development of Electrochemical Cells

Another approach that is taken with regard to the electrification of the ammonia process is through the development of the electrochemical HB (eHB) process. The recent article published by Kyriakou et al. proposed the implementation of BaZrO3-based protonic ceramic membrane reactor (PCMR). In a simple explanation, PCMR integrated key steps of the HB process of SMR, WGS, and ammonia synthesis into a single cell. Suggested PCMR operates at atmospheric pressure and temperature of 550 °C–650 °C that is significantly less than the temperature and pressure requirement of conventional HB Process ($T = 800$ °C ~ 1000 °C & $P > 100$ bar). Figure 4.9 shows the general overview on how PCMR functions [44].

Driven by renewable energy, the CH_4-H_2O mixture goes through SMR and WGS reactions on Ni anode to produce CO_2, H+, and electrons. The protons are transported across the BZCY81 membrane and react with lattice nitrogen on cathode that uses the VN-Fe electrocatalyst. The decrease in hydrogen in the anode chamber tilts the equilibrium of SMR and WGS reactions, converting most of the methane to CO_2 (>95%). Subsequently, through a series of testing of a system that coupled PCMR with a protonic ceramic fuel cell (PCFC) achieved a maximum ammonia yield rate of $115.8\frac{\mu g}{cm^2 * h}$ and FE of 5.5% under the condition of 6.3 V and 600. Further explanation shows the result that FE above 35% is the minimum threshold for the proposed eHB system to be comparable to an industrial plant [44].

The significance of this research, however, comes from the successful emulation of thermodynamic reaction via the electrochemical method. It also presents itself to be in a much more practical position than other comparable research that is done in the research topic of electrification of ammonia synthesis. The integration of a high-cost compartment into a single cell represents the possibility of scaling down the ammonia processing plant that opens its possibility in decentralized production of ammonia that will further reduce the carbon footprint resulting from the ammonia transport. The optimization of the system and integration of improving catalysts should provide a constructive outlook on the concept of the eHB system.

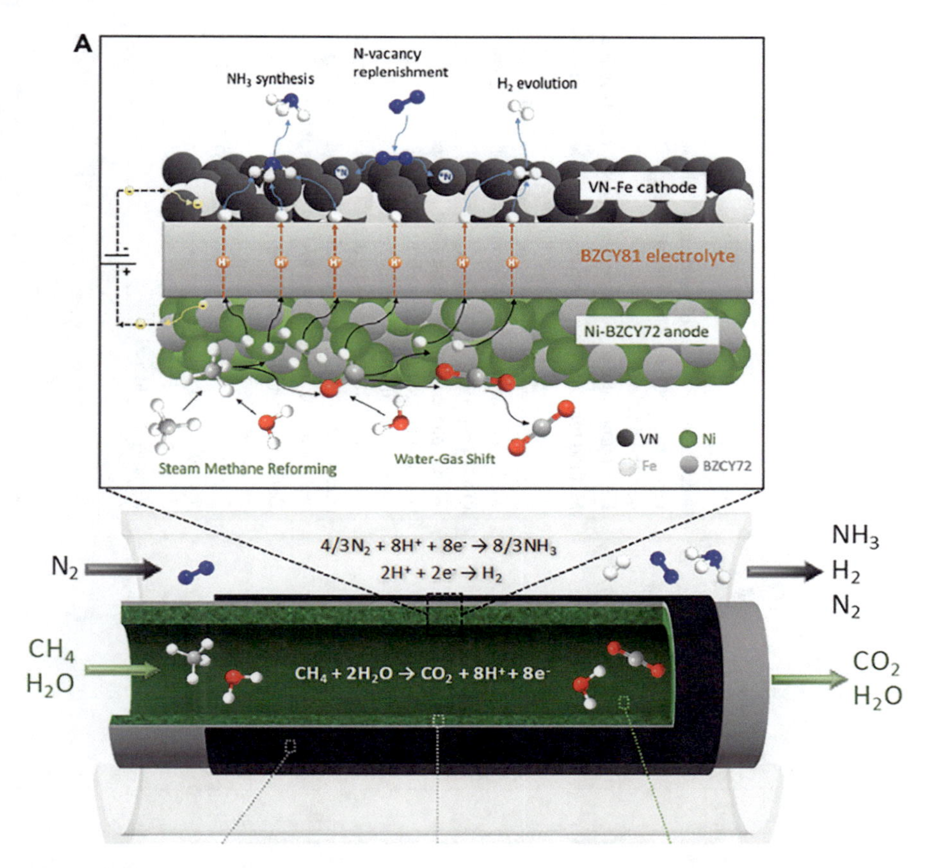

Fig. 4.9 Schematic of protonoic ceramic membrane reactor (PCMR) and general overview of methodology

4.5 Classification and Discussion

4.5.1 Power-to-Heat

The following chart (Table 4.3) represents the papers reviewed for the Power-to-Heat section of this report.

All the sources are relatively recent, with each source being published within the past 3 years. This goes to show how relevant the topic is to society today.

The initial two studies discussed in the Power-to-Heat section of this report serve to introduce the topic and describe how it operates. These sources are followed by a source that discusses the positive impacts of Power-to-Heat systems, including energy savings and emissions reduction. The next source discussed identifies a way to a couple of local district heating and electrical distribution grids. This provides an innovative way to use the Power-to-Heat system to efficiently improve

Table 4.3 Classification of Power-to-Heat-related literatures

References	Study	Research topic	Scale	Region	Publication year	Current state of research	Highlights
[6]	P2H	Technologies, modeling	Residential	NA	2018	Early	-odel-based analyses of residential P2H options Fossil fuel substitution, renewable integration, decarbonization Heat pumps and passive thermal storage are favorable
[7]	P2H	District heating	Local	NA	2019	Early	Method that enables a detailed technical assessment of the operation of coupled heat and power networks Different use cases where a local coupling of the networks with power-to-heat is supporting the transition to smart energy networks
[8]	P2H	Demand flexibility	Residential	NA	2018	Early	Water, phase change material, and thermochemical material tanks are integrated for optimal control Demand flexibility of thermal energy storage tanks integrated with a building heating system is quantified Flexibility indicators representing demand flexibility are calculated for reference and optimal control
[9]	P2H	Regionalized eat demand in Germany	Residential	DE, EU	2020	Early	Determination of heat demand share covered by electric heating technologies 729 building categories defined from a special evaluation of census data Five classes of heating types and installed heating capacity considered

local heating and electricity networks. The second to last source reviewed in the Power-to-Heat section of this report identifies an area of improvement for Power-to-Heat options: flexibility of Power-to-Heat and thermal energy storage when it comes to building heating systems. The final source reviewed for Power-to-Heat provides a case study in Germany for implementing Power-to-Heat systems in each administrative district in Germany. This is an example of how Power-to-Heat systems can be implemented into communities.

The papers reviewed for the Power-to-Heat topic all had similar outlooks for the future of these systems. They were all optimistic about its future applications and success within various heating applications. Future research into Power-to-Heat systems should consider evolving electricity sources as well as evolving heating systems. Future studies may also need to consider thermal energy storage systems that support flexibility in heating demands (Tables 4.4 and 4.5).

4.5.2 Power-to-Gas

For the study on the Power-to-Gas-related topic, the literature classification is mainly broken down into two major research topics: the electrolysis of water and the compressed-air energy systems, which correspond to the chemical approach and the potential approach, respectively. For this, the study is mainly focused on articles dated from 2014 to 2015, with the exception of one article in 2012 to introduce one of the first few studies that applied a multi-physics model that takes into account various parameters to describe the electrolyte system. The region of research on the topics centered around the European Union and China. Not many studies on the modeling were industrial-scale electrolyte systems; therefore, the literature review paper focused only on lab-scale modeling of electrolyte systems. Most of the articles are still in the early stage of research due to the novelty of the topic. In the case of compressed-air energy systems, we focused on some comparative analysis on different systems and innovative approaches that could be made to increase the system efficiency.

4.5.3 Power-to-Chemicals

Eleven literatures were discussed for Power-to-Chemicals. Except for two articles that either provided a general basis on the ammonia synthesis or discussed innovative development of electrochemical PCMR cells, the majority of the articles discussed the experimental research on electrocatalysts or the computational method of catalyst research. To provide recent information on the project topic, the emphasis was given to select those that were published recently. As a result, 7 out of 11 chosen articles are published between 2019 and 2020. The research papers on electrocatalysts utilized product chemical yield rate and Faradaic Efficiency to

Table 4.4 Classification of Power-to-Gas-related literatures

References	Study	Research topic	Scale	Region	Publication year	Current state of research	Highlights
[18]	P2G	Electrolysis of water	Industrial	CN	2019	Early	Alkaline water electrolysis with a flow rate of 1000 m^3/h has been achieved whereas in the case of proton exchange membrane technology, the most advanced equipment could produce 400m^3/h of hydrogen. Alkaline water electrolysis showed downsides such as complex maintenance due to corrosion and slow start-up. Meanwhile, proton exchange membrane technology requires a much higher expenditure.
[19]	P2G	Electrolysis of water	Industrial	CN	2019	Early	Compared the AWE and the PEM and on top of that a high-temperature steam electrolysis. Results showed a higher efficiency of water decomposition at higher temperature.
[20]	P2G	Electrolysis of water	Industrial	EU	2019	Early	Compared PEM water electrolysis and steam methane reforming through life cycle assessment. Results suggested that the PEM method is capable of reducing CO_2 emission by 75% and global warming potential to 3.3 kg CO_2.
[21]	P2G	Electrolysis of water	Industrial	CA	2012	Intermediate	Developed a multi-physics model by taking into account all variations of structural parameters unlike most of the conventional approaches which only take into account the thermal aspect. This approach can be applied in a wide range of alkaline electrolyzers, and one of the highlights of this approach is that the characterization can be completed in a relatively short time.
[22]	P2G	Electrolysis of water	Lab	EU	2014	Early	Developed a physics model (alkaline electrolyzer simulation tool AEST) to describe the electrodes' physical and electrical performance and estimate the optimal operating current.

[23]	P2G	Electrolysis of water	Semi-industrial	EU	2016	Early	Developed a model under bond graph formalism which allows accurate description of the dynamic behavior of a semi-industrial PEM electrolysis (25 kW).
[27]	P2G	Compressed-air energy systems	Industrial	CN	2018	Early	Conducted a comparison analysis on the different CAES systems. The study showed that advanced A-CAES has a better efficiency compared to the conventional CAES which has larger generating capacity at the cost of much waste heat and a lower system efficiency.
[28]	P2G	Compressed-air energy systems	Industrial	CN	2017	Early	Proposed an improved version of the conventional CAES system by applying a pre-cooling system to decrease the energy consumption. The RTE was proved to increase by 3% compared to the conventional CAES systems.
[29]	P2G	Compressed-air energy systems	Industrial	AUS	2020	Early	The study showed that a higher compressor flow rate reduces the system efficiency and increasing the expander flow rate reduces the system and heat recovery efficiencies up to 8.8%–40.3%,

Table 4.5 Classification of Power-to-Chemical-related literatures

References	Study	Research topic	Scale	Region	Publication year	Current state of research	Highlights
[32]	P2C	Decarbonization of ammonia	N/A	UK,EU	2017	Review	Strategies in decarbonization in the ammonia processing industry Highlights discrepancy between requirement and current research progress Techno-economic evaluation
[34]	P2C	Electrocatalysts	Lab	CN	2020	Early	Collective analysis on the research on the electrochemical synthesis of ammonia by nitrogen reduction reaction
[35]	P2C	Electrocatalysts	Lab	US/JP	2013	Early	The importance of catalyst design, electrolyte choice, and electrode structure The biggest opportunities for performance enhancement are highlighted
[36]	P2C	Computational electrocatalyst	Lab	US	2016	Intermediate	Studies on scaling relationship of binding energies among the catalytic intermediates
[37]	P2C	Electrocatalysts	Lab	US/CN	2020	Early	Discusses the CO_2 reduction strategies that aim to improve the performance of catalyst by decoupling the scaling relationship
[38]	P2C	Electrocatalysts	Lab	CN	2020	Early	Introduction of silicotungstic acid as an electron carrier which was found to improve the hydrogen yield by 35.31% Identified 10.5 and 13.3 times improvement for ammonia production rate and FE for Pt/C catalyst
[40]	P2C	Computational electrocatalyst	Lab	CA	2020	Early	Developed DFT trained NN ML model that exerted high accuracy in HEA electrocatalyst research Model is highly generalizable to other bimetallic catalysts while maintaining reasonable simplicity
[41]	P2C	Computational electrocatalyst	Lab	KR	2019	Early	Utilized slab graphical convolutional neural network (SGCNN) method machine learning model to identify catalysts with low AE and high FE in relative to base catalyst.

[42]	P2C	Computational electrocatalyst	Lab	Pakistan	2020	Early	Using DNN, successfully evaluated the efficiency of B-doped graphene SACs Reduced computation time by implementing and algorithm to remove insignificant catalyst from screening
[43]	P2C	Computational electrocatalyst	Lab	CN	2020	Early	Developed the new machine learning method that incorporated 315 C1/C2 intermediate and transition states apparent in the reaction pathway from syngas to ethanol Integrated three recently proposed training methods of graph convolutions, weave, and graph neural network in a study
[44]	P2C	Electrochemical cells	Lab	GR,EU	2020	Intermediate	Developed BaZrO3-based PCMR that incorporated important function of WGS, SMR, and ammonia synthesis into a single cell Operates at ambient pressure and temperature (400 ~ 500C) that is lower than what is required for the traditional HB process

judge the performance and relative success of the experiment. Those that discussed computational studies on catalysts focused on various ways that ML can benefit the catalysts' research in terms of cost reduction, time efficiency, and new methodologies. Both the experimental and computational catalyst researches are in an infant stage of research as the results are still significantly worse than the minimum recommended requirement for technology to be practical. One literature that proposed the development of new electrochemical cells was very different from other literature that was discussed in this section but suggested an alternative method that possesses the high potential to come to fruition in the relatively near future. The regional distribution of literature is well distributed around the world, but the trend that was observed from the research phase indicated that Europe, China, and the United States are three regions that are currently performing a lot of research in this field.

4.6 Conclusion

The fundamental basis of Power-to-X that aims to reduce carbon emission by providing alternative options to fossil fuels is crucial as we try to mitigate climate change. Recently observed economic competitiveness of renewable energies provides sufficient motive for researchers and countries to research in the technology that is both competent and environmentally friendly. This survey report focused on identifying the recent progress on three key categories of PtX which: Power-to-Heat, Power-to-Gas, and Power-to-Chemicals. While PtX technically does not seem to be the popular terminology for researchers out of Europe, a lot of studies are performed on various methods that aim to decarbonize chemical processing industries through electrification. Different timelines are drawn for each of PtH, PtG, and PtC technologies. PtH technology is the first one to come to fruition as the successful implementation of technologies like an electric heat pump in the residential area leaves PtH with optimization and integration problems that would have to be configured. PtG technology is the one to follow as there are numerous small-scale projects already in full operation, leaving the obstacle of facing technical issues associated with scaling up. PtC technology is still in a very early stage of the research which will require a lot of time before we see a large-scale PtC plant in operation. Regardless, research literature that is high in quality and quantity is going to provide answers that will help in the journey to net-zero carbon emission.

References

1. Siemens, HYFLEXPOWER: The world's first integrated power-to-X-to-power hy (2020, May 29). Retrieved from https://press.siemens.com/global/en/pressrelease/hyflexpower-worlds-first-integrated-power-x-power-hydrogen-gas-turbine-demonstrator

2. M. De Graff, Electrification of the chemical industry (n.d.). Retrieved from https://www.tno.nl/en/focus-areas/industry/roadmaps/sustainable-chemical-industry/voltachem-for-electrification-in-the-chemical-industry/electrification-of-the-chemical-industry/
3. Berenschot, CE Delft, Industrial Energy Experts, and Energy Matters, Electrification in the Dutch process industry [White paper] (2017, February 8). Retrieved from https://cedelft.eu/publications/electrification-in-the-dutch-process-industry
4. K. Kranenburg, E. Schols, H. Gelevert, R. Kler, Y. Delft, and M. Weeda, Empowering the chemical industries: Opportunities for electrification [White paper]. TNO (2016). https://www.tno.nl/media/7514/voltachem_electrification_whitepaper_2016.pdf
5. International Aspects of a Power-to-X Roadmap, (2018, October 18). Retrieved from: https://www.frontier-economics.com/media/2642/frontier-int-ptx-roadmap-stc-12-10-18-final-report.pdf
6. A. Bloess, W.P. Schill, A. Zerrahn, Power-to-heat for renewable energy integration: A review of technologies, modeling approaches, and flexibility potentials. Appl. Energy **212**, 1611–1626 (2018)
7. B. Leitner, E. Widl, W. Gawlik, R. Hofmann, A method for technical assessment of power-to-heat use cases to couple local district heating and electrical distribution grids. Energy **182**, 729–738 (2019)
8. C. Finck, R. Li, R. Kramer, W. Zeiler, Quantifying demand flexibility of power-to-heat and thermal energy storage in the control of building heating systems. Appl. Energy **209**, 409–425 (2018)
9. W. Heitkoetter, W. Medjroubi, T. Vogt, C. Agert, Regionalised heat demand and power-to-heat capacities in Germany – An open dataset for assessing renewable energy integration. Appl. Energy **259**, 114161 (2020)
10. M. Thema, F. Bauer, M. Sterner, Power-to-gas: Electrolysis and methanation status review. Renew. Sust. Energ. Rev. **112**, 775–787 (2019). https://doi.org/10.1016/j.rser.2019.06.030
11. Integrated Power-to-Gas and Gas-to-Power with Air and Natural-Gas Storage, (n.d.). doi: https://doi.org/10.1021/acs.iecr.8b04711.s001
12. R.Y. Kannah, S. Kavitha, O.P. Karthikeyan, G. Kumar, N.V. Dai-Viet, J.R. Banu, Techno-economic assessment of various hydrogen production methods – A review. Bioresour. Technol. **319**, 124175 (2021). https://doi.org/10.1016/j.biortech.2020.124175
13. S. Grigoriev, V. Fateev, D. Bessarabov, P. Millet, Current status, research trends, and challenges in water electrolysis science and technology. Int. J. Hydrog. Energy **45**(49), 26036–26058 (2020). https://doi.org/10.1016/j.ijhydene.2020.03.109
14. C. Wulf, J. Linßen, P. Zapp, Review of power-to-gas projects in Europe. Energy Procedia **155**, 367–378 (2018). https://doi.org/10.1016/j.egypro.2018.11.041
15. GreenCarCongress, Shell, ITM Power to install 10MW electrolyzer for refinery hydrogen (2017, September 01). Retrieved from http://www.greencarcongress.com/2017/09/20170901-shell.html
16. M. Götz, J. Lefebvre, F. Mörs, A.M. Koch, F. Graf, S. Bajohr, et al., Renewable power-to-gas: A technological and economic review. Renew. Energy **85**, 1371–1390 (2016). https://doi.org/10.1016/j.renene.2015.07.066
17. GreenCarCongress, Thyssenkrupp offering large-scale water electrolysis hydrogen (2018, July 17). Retrieved from http://www.greencarcongress.com/2017/09/20170901-shell.html
18. Y. Guo, G. Li, J. Zhou, Y. Liu, Comparison between hydrogen production by alkaline water electrolysis and hydrogen production by PEM electrolysis. IOP Conf. Ser.: Earth Environ. Sci. **371**, 042022 (2019). https://doi.org/10.1088/1755-1315/371/4/042022
19. A. Mazza, E. Bompard, G. Chicco, Applications of power to gas technologies in emerging electrical systems. Renew. Sust. Energ. Rev. **92**, 794–806 (2018). https://doi.org/10.1016/j.rser.2018.04.072
20. K. Bareiß, C.D. Rua, M. Möckl, T. Hamacher, Life cycle assessment of hydrogen from proton exchange membrane water electrolysis in future energy systems. Appl. Energy **237**, 862–872 (2019). https://doi.org/10.1016/j.apenergy.2019.01.001

21. M. Hammoudi, C. Henao, K. Agbossou, Y. Dubé, M. Doumbia, New multi-physics approach for modelling and design of alkaline electrolyzers. Int. J. Hydrog. Energy **37**(19), 13895–13913 (2012). https://doi.org/10.1016/j.ijhydene.2012.07.015
22. C. Henao, K. Agbossou, M. Hammoudi, Y. Dubé, A. Cardenas, Simulation tool based on a physics model and an electrical analogy for an alkaline electrolyser. J. Power Sources **250**, 58–67 (2014). https://doi.org/10.1016/j.jpowsour.2013.10.086
23. P. Olivier, C. Bourasseau, B. Bouamama, Modelling, simulation and analysis of a PEM electrolysis system. IFAC-PapersOnLine **49**(12), 1014–1019 (2016). https://doi.org/10.1016/j.ifacol.2016.07.575
24. EASE/EERA, European energy storage technology development roadmap [White Paper] (2017). Retrieved from https://eera-es.eu/wp-content/uploads/2016/03/EASE-EERA-Storage-Technology-Development- Roadmap-2017-HR.pdf
25. A.J. Giramonti, R.D. Lessard, W.A. Blecher, E.B. Smith, Conceptual design of compressed air energy storage electric power systems. Appl. Energy **4**(4), 231–249 (1978). https://doi.org/10.1016/0306-2619(78)90023-5
26. S.D. Garvey, Compressed air energy storage: underground technologies [PowerPoint slides] (2019). Retrieved from https://energnet.eu/sites/default/files/1-Garvey_at_EWUES_3_for%20web.pdf
27. S. Zhou, J. Zhang, W. Song, Z. Feng, Comparison analysis of different compressed air energy storage systems. Energy Procedia **152**, 162–167 (2018). https://doi.org/10.1016/j.egypro.2018.09.075
28. L. Chen, P. Hu, C. Sheng, M. Xie, A novel compressed air energy storage (CAES) system combined with pre-cooler and using low grade waste heat as heat source. Energy **131**, 259–266 (2017). https://doi.org/10.1016/j.energy.2017.05.047
29. H. Mozayeni, X. Wang, M. Negnevitsky, Dynamic analysis of a low-temperature adiabatic compressed air energy storage system. J. Clean. Prod. **276**, 124323 (2020). https://doi.org/10.1016/j.jclepro.2020.124323
30. cefic, 2020 Facts & Figures of the European chemical industry [White paper] (2020). Retrieved from https://cefic.org/app/uploads/2019/01/The-European-Chemical-Industry-Facts-And-Figures-2020.pdf
31. Z.J. Schiffer, K. Manthiram, Electrification and Decarbonization of the chemical industry. Joule **1**(1), 10–14 (2017). https://doi.org/10.1016/j.joule.2017.07.008
32. L. Ye, R. Nayak-Luke, R. Bañares-Alcántara, E. Tsang, Reaction: "Green" ammonia production. Chem **3**(5), 712–714 (2017). https://doi.org/10.1016/j.chempr.2017.10.016
33. H. Xu, K. Ithisuphalap, Y. Li, S. Mukherjee, J. Lattimer, G. Soloveichik, G. Wu, Electrochemical ammonia synthesis through N2 and H2O under ambient conditions: Theory, practices, and challenges for catalysts and electrolytes. Nano Energy **69**, 104469 (2020). https://doi.org/10.1016/j.nanoen.2020.104469
34. T. Wu, W. Fan, Y. Zhang, F. Zhang, Electrochemical synthesis of ammonia: Progress and challenges. Mater. Today Phys. **100310** (2020). https://doi.org/10.1016/j.mtphys.2020.100310
35. H. Jhong, S. Ma, P.J. Kenis, Electrochemical conversion of CO_2 to useful chemicals: Current status, remaining challenges, and future opportunities. Curr. Opin. Chem. Eng. **2**(2), 191–199 (2013). https://doi.org/10.1016/j.coche.2013.03.005
36. J. Greeley, Theoretical heterogeneous catalysis: Scaling relationships and computational catalyst design. Annu. Rev. Chem. Biomol. Eng. **7**(1), 605–635 (2016). https://doi.org/10.1146/annurev-chembioeng-080615-034413
37. L. Fan, C. Xia, F. Yang, J. Wang, H. Wang, Y. Lu, Strategies in catalysts and electrolyzer design for electrochemical CO_2 reduction toward C2 products. Science. Advances **6**(8) (2020). https://doi.org/10.1126/sciadv.aay3111
38. Y. Zhao, T. Wang, H. Wang, S. Lu, Y. Wang, M. Zhao, et al., Study on POM assisted electrolysis for hydrogen and ammonia production. Int. J. Hydrog. Energy **45**(53), 28313–28324 (2020). https://doi.org/10.1016/j.ijhydene.2020.07.147

39. W. Thiel, Computational catalysis-past, present, and future. Angew. Chem. Int. Ed. **53**(33), 8605–8613 (2014). https://doi.org/10.1002/anie.201402118
40. Z. Lu, Z.W. Chen, C.V. Singh, Neural network-assisted development of high-entropy alloy catalysts: Decoupling ligand and coordination effects. Matter **3**(4), 1318–1333 (2020). https://doi.org/10.1016/j.matt.2020.07.029
41. M. Kim, B.C. Yeo, Y. Park, H.M. Lee, S.S. Han, D. Kim, Artificial intelligence to accelerate the discovery of N2 Electroreduction catalysts. Chem. Mater. **32**(2), 709–720 (2019). https://doi.org/10.1021/acs.chemmater.9b03686
42. M. Zafari, D. Kumar, M. Umer, K.S. Kim, Machine learning-based high throughput screening for nitrogen fixation on boron-doped single atom catalysts. J. Mater. Chem. A **8**(10), 5209–5216 (2020). https://doi.org/10.1039/c9ta12608b
43. B. Wang, T. Gu, Y. Lu, B. Yang, Prediction of energies for reaction intermediates and transition states on catalyst surfaces using graph-based machine learning models. Mol. Catal. **498**, 111266 (2020). https://doi.org/10.1016/j.mcat.2020.111266
44. V. Kyriakou, I. Garagounis, A. Vourros, E. Vasileiou, M. Stoukides, An electrochemical Haber-Bosch process. Joule **4**(1), 142–158 (2020). https://doi.org/10.1016/j.joule.2019.10.006

Chapter 5
Machine Learning Models for Absorption-Based Post-combustion Carbon Capture

Fatima Ghiasi, Ali Ahmadian, Kourosh Zanganeh, Ahmed Shafeen, and Ali Elkamel

5.1 Introduction

Over the last hundred years, the average global temperature has been rising. Research in the field has strongly linked the shift in climate, particularly the rise in average global temperature, to rising atmospheric CO_2 concentrations [1]. During this past century, CO_2 concentrations have risen from 301 ppm to 408 ppm [2]. These levels are projected to continue to increase if the contemporary policies and technologies are maintained [3]. For this reason, it is important to limit future carbon emissions.

One method of reducing future emissions is to substitute current industrial processes that emit greenhouse gases with less carbon-intensive alternatives. However, replacing existing infrastructure is costly and is only feasible as a long-term strategy. In addition, many industrial processes that produce a large amount of

F. Ghiasi
Department of Systems Design Engineering, University of Waterloo, Waterloo, ON, Canada
e-mail: fghiasi@uwaterloo.ca

A. Ahmadian (✉)
Department of Electrical Engineering, University of Bonab, Bonab, Iran

Department of Chemical Engineering, University of Waterloo, Waterloo, ON, Canada
e-mail: ahmadian@uwaterloo.ca

K. Zanganeh · A. Shafeen
Natural Resources Canada (NRCan), Canmet ENERGY-Ottawa (CE-O), Ottawa, ON, Canada
e-mail: kourosh.zanganeh@NRCan-RNCan.gc.ca; ahmed.shafeen@NRCan-RNCan.gc.ca

A. Elkamel
Department of Chemical Engineering, University of Waterloo, Waterloo, ON, Canada

Department of Chemical Engineering, Khalifa University, Abu Dhabi, United Arab Emirates
e-mail: aelkamel@uwaterloo.ca

A. Ahmadian et al. (eds.), *Carbon Capture, Utilization, and Storage Technologies*, Green Energy and Technology, https://doi.org/10.1007/978-3-031-46590-1_5

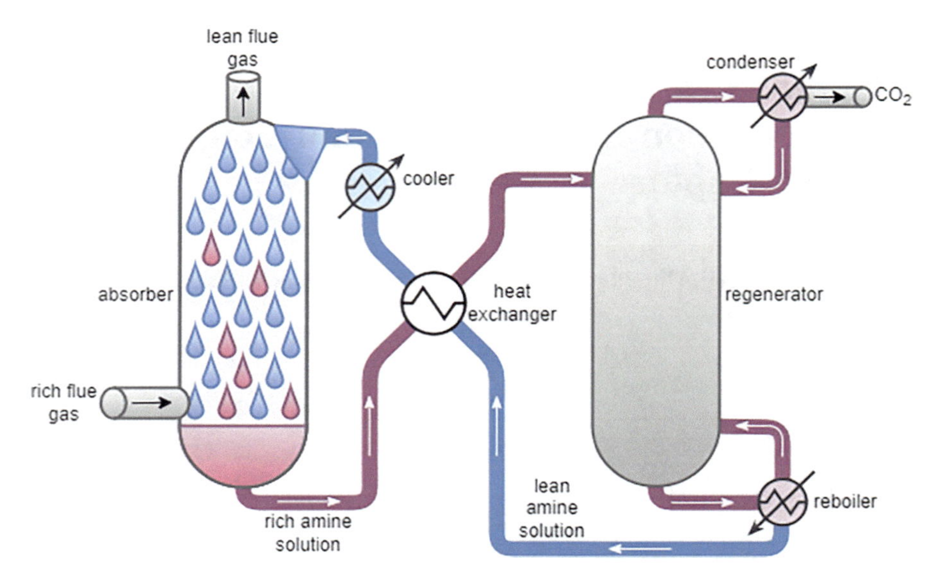

Fig. 5.1 Process of amine flue gas scrubbing

carbon dioxide as a by-product, such as conventional steelmaking and cement production, cannot be easily replaced. Approximately 46% of global greenhouse gases are emitted from stationary point sources [4]. For stationary point sources, one way of lowering carbon emissions is to divert produced carbon dioxide away from the atmosphere, in a process called post-combustion carbon capture (PCCC).

The flue gas emitted by industrial plants contains carbon dioxide mixed with a large quantity of other gases such as nitrogen, oxygen, and water vapor. However, many applications that use CO_2 require the purity of the gas be above a certain threshold. Additionally, the transport and storage of these extra gases requires energy and space. For these reasons, it is important to use a method to separate the CO_2 from other flue gas constituents.

A configuration which can be easily implemented is to append a separate post-combustion carbon capture unit to an already existing plant. The most mature carbon capture technology is amine scrubbing, which has been used to remove carbon dioxide from natural gas since the 1930s [5]. The amine scrubbing process is shown below in Fig. 5.1.

The rich flue gas is pumped from the bottom of the absorber. As it flows upwards, the carbon dioxide within the flue gas is absorbed by the dripping lean amine solution. The rich solution then crosses a heat exchanger and enters the regenerator. At the bottom of the regenerator, the reboiler heats the rich solution, evaporating the CO_2 along with small amounts of the solution. The liquid that remains in the reboiler is the lean amine solution, which crosses a heat exchanger and is cooled before being pumped to the top of the absorber column. At the top of the regenerator, the condenser cools the vapor, condensing and recovering the amine solution. The gas

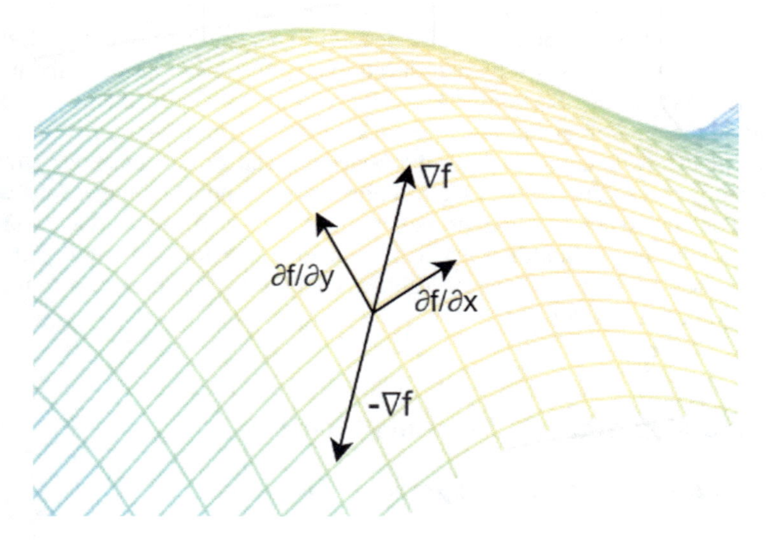

Fig. 5.2 Partial derivatives, gradient, and negative gradient of a function

that remains in the condenser is the CO_2, which is then pumped away for storage or use.

The major barrier to adopting post-combustion carbon capture units is that they are not economically feasible due to their high energy requirements. If carbon capture units can be optimized to require a lower energy input, more industries will be inclined to implement this technology. One method of optimizing a system is to use simulations. Different sets of parameters are tested in different instances of the simulation, and the best performing set is chosen to be implemented in building prototypes. However, before using this method, models that can accurately predict the behavior of the system in question must be created.

There are two main methods of developing models, mechanistic and empirical. Mechanistic models are built from the ground up using theoretical relationships between the fundamental components of the system. They are suitable for systems in which all components are known, and their interactions are well-defined. However, for many systems in real life, the internal mechanisms cannot be seen or inferred. For these cases, empirical modeling should be used.

Empirical models are based on observed experimental data. They do not require extensive knowledge of system fundamentals. One large subset of empirical modelling is regression. In regression, the goal is to map independent variables, which are defined outside of the model, to dependent variables. In machine learning, algorithms are used to aid in generating regression models. Machine learning models can be as straightforward as linear regression or as opaque as ensemble learning models. The quality of a machine learning model is limited by both the quality of the data used to create it and the quality of the process used to generate the model.

Figure 5.2 displays a chart of all the topics that will be covered in this chapter. In Sections 5.2 and 5.3, general machine learning concepts and relevant classes of machine learning models will be introduced to help the reader evaluate the individual models themselves. Sections 5.4 and 5.5 will explore applications of machine learning in the field of amine-based post-combustion carbon capture.

This chapter will highlight both the beneficial and disadvantageous strategies employed by researchers developing machine learning models. Suggestions will be provided on how to potentially improve certain models. It will serve as a resource for future researchers, highlighting efficient strategies and how to avoid pitfalls of strategies used by past researchers. Table 5.1 represents the main topics that will be covered in this chapter.

5.2 Training Machine Learning Models

5.2.1 Supervised and Unsupervised Learning

In machine learning, learning can either be supervised or unsupervised. In supervised learning, pairs of input and output variables are given to a model. The goal of training is to make the model able to predict output values based on a given set of input values. Supervised learning can be regression, where the output is a numerical value, or classification, where the output is a label. Contrastingly, in unsupervised learning, only input values are given. The job of the machine learning algorithm is to infer similarities about different data points. It is commonly used to discover patterns between variables and cluster data points into unlabeled groups. Unsupervised learning has limited usefulness in the field of carbon capture. However, it is useful for generating vector numerical descriptors of entities, such as molecular structures [6]. These vector numerical descriptions can then be used as inputs for supervised learning algorithms.

5.2.2 Quantifying Performance of Supervised Models

When training a supervised model, a measure of performance is necessary. The quality of a model is based on how closely it can predict a resultant value based on a set of given parameters. For quantifying how well the model describes the data, machine learning algorithms use a function called the loss function. There are different loss functions, and each is useful for a different type of task. Common loss functions include the mean absolute error (MAE), the mean square error (MSE), and cross entropy loss. These different loss functions are compared in Table 5.2. In supervised machine learning, the training algorithm, the optimizer, strives to minimize the loss function for a set of training points by changing parameters of the model. The most commonly utilized optimization algorithms are gradient based.

Table 5.1 Topics that will be covered in this chapter

Concepts							
General training methods			Available models				
Quantifying performance	Gradient descent	Avoiding pitfalls of training models	K nearest neighbors	Support vector machines	Neural networks	Fuzzy logic	Tree-based methods
Applications							
Predicting solution properties					Modelling CO_2 capture systems		
Mechanical properties	CO_2 solubility	Reaction rate	Environmental toxicity	Lipophilicity	CO_2 capture level	Power requirements	Pollution release

Table 5.2 Common loss functions used in machine learning

Loss function	Mean absolute error (MAE)	Mean square error (MSE)	Cross entropy loss (CEL)
Formula	$\text{MAE} = \frac{1}{n}\sum_{i=1}^{n} \lvert y_i - \widehat{y}_i \rvert$	$\text{MSE} = \frac{1}{n}\sum_{i=1}^{n} (y_i - \widehat{y}_i)^2$	$\text{CEL} = -\frac{1}{n}\sum_{i=1}^{n}\sum_{j=1}^{m} y_{ij}\log(p_{ij})$
Tasks that are useful for:	Regression	Regression	Classification
Benefits	Less sensitive to outliers	Larger mistakes contribute more to loss	Strongly disincentivizes misclassification

Fig. 5.3 Training and test data performance with and without overfitting

5.2.3 Gradient Descent

Gradient descent is a method to find the minimum of a function. In gradient descent, the values of the variables decrease by an amount proportional to their partial derivative. The direction of the next point is the opposite of the direction of the gradient. It can be thought of as climbing down a hill. The partial derivatives, the gradient, and the negative gradient of an arbitrary curve are shown in Fig. 5.3.

In classic gradient descent, the machine learning algorithm calculates the loss function and gradient based on all of the data points present within the training data set. Then, it adjusts the parameters of the ML model by subtracting the gradient scaled by the training rate. This method is particularly slow for larger datasets, since the parameters can only be updated once every training data point has been analyzed. An alternative to this method is Stochastic Gradient Descent (SGD). In SGD, the loss function and gradient are estimated based on a subset of the training data points, called a batch. Based on the gradient of the loss function for the batch, the parameters of the model are updated. Since the algorithm only needs to look at a subset of data points before updating parameters, they are adjusted more frequently. This leads to faster convergence.

Many of the most commonly used optimization algorithms are derivatives of SGD. They aim to provide more rapid convergence. One popular class of SGD derivatives are momentum-based optimization methods, such as Adam [7]. In momentum-based methods, the gradient causes the momentum to increase by an amount proportional to it. The parameter vector's position is changed based on the momentum. In addition to having faster convergence, momentum-based optimizers do not get stuck at saddle points, since they do not rely solely on the local gradient. One pitfall of gradient descent is that if there exist local maxima or minima, the optimizer risks becoming stuck around those local peaks and troughs. There are different methods of mitigating the risk of falling into local minima. The most common way is to train multiple instances of the model in parallel. If the initial guess of each instance is randomly distributed within the solution space, the probability of at least one of the instances falling toward the global minimum is greater than the probability of each individual instance falling toward the global minimum.

5.2.4 *Pitfalls and Mitigation Techniques*

When training a model repeatedly, it may learn to memorize individual data points instead of learning general trends. This is called overfitting. Machine learning models that are too complex for a given problem are prone to this. For this reason, the performance is evaluated on a set of data points not used for training, the test set. If a model shows good performance for the training set but poor performance for the test set, it indicates that it is overfitting and failing to generalize. This is shown in Fig. 5.3.

The opposite problem is underfitting, where the model does not have enough training parameters to represent the relationships between variables. If a model shows poor performance for both training and test data, it is likely underfitting. There exist algorithms for determining the optimal number of parameters and structure of a machine learning model, called hyperparameter optimization. A common, but inefficient, hyperparameter optimization algorithm, called grid search, involves training models with every combination of possible hyperparameters, and choosing the one that has the best performance. Additionally, one potential problem that may arise is that a given training set may not be representative of the system. This would lead the model to have poor performance regardless of the amount of training. To mitigate this problem, multiple instances of a model can be trained with different points assigned to the train and test sets. This is called k-fold cross validation.

In ensemble learning methods, multiple instances of a model are trained in parallel, each with different training sets. The output of the consensus model is the aggregate of the outputs of each instance. For classification problems, it is the most common predicted class. For regression problems, it is the mean output value. For ensemble learning methods, each individual instance has to be trained on an independent set of training points. One method of generating a diverse array of training sets is bootstrap aggregation. In bootstrap aggregation, the training set for a particular instance is chosen, with replacement, from a bank of potential training points. Some points may be chosen more than once, while others may not be chosen.

5.2.4.1 Data Pre-processing

Raw data cannot always be used directly with machine learning models. For example, if time series data contains a large amount of noise or missing values, the training algorithm will be unable to perform well. One common method of suppressing noise is to take multiple measurements and record the mean value. For data that is not time sensitive, such as equilibrium chemical compositions, measurement devices can take multiple time consecutive readings from the same device and store or display the average. However, for time-sensitive data, such as time series readings, multiple sensors would need to be used to suppress noise using this method. As an alternative to using multiple sensors, in many cases, noise is removed from time series data by post-processing algorithms. These algorithms can be as simple as digital filters, to as complex as deep learning convolutional neural networks [8]. In addition, simple interpolation algorithms or machine learning models can also infer the value of missing measurements [9].

If certain machine learning models, such as neural networks, have inputs that are orders of magnitude different in numerical values, they might prioritize the variable with the larger number as being more important. To avoid this situation, features are scaled to have number values that are similar in magnitude. Furthermore, if the input data is not homogenously distributed across the input space, as shown in Fig. 5.4, there may exist regions of the input space, indicated by red triangles, that have sparse or nonexistent training points. The machine learning model should not be expected to perform well in those regions. In addition, if two inputs appear to have a correlation, the machine learning model may interpret effects due to one variable as being due to another. It is best practice to use a training data set where the inputs are independent of each other.

However, in some cases, this is not possible. In this case, two new dimensions should be created based on the two inputs, as shown in the right graph of Fig. 5.4. One dimension, the principal component u_1, should be parallel to the apparent trend

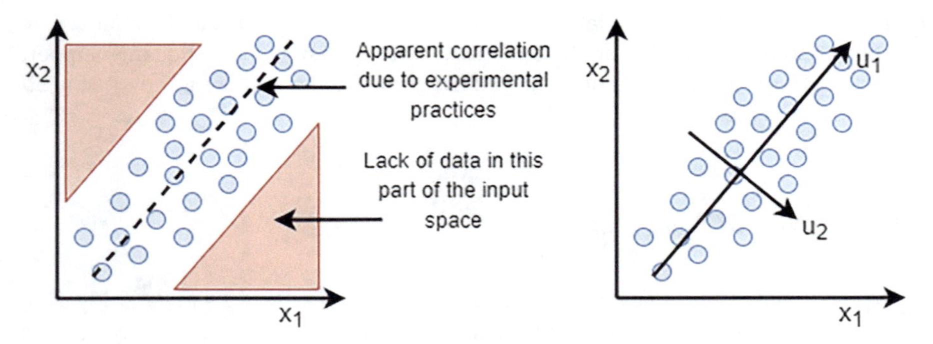

Fig. 5.4 Principal component analysis

that exists between the two variables. Another dimension, u_2, should be orthogonal to the first dimension and serve as a measure of deviation from the trend.

5.3 Available Machine Learning Models

This chapter covers the types of machine learning models that have been used to model processes or predict outcomes within absorption-based post-combustion carbon capture.

5.3.1 k-Nearest Neighbors

K-nearest neighbors (KNN) is the most straightforward machine learning algorithm. In kNN, the training data set is stored in memory. When an input is given to the system, it scrolls through the data points in memory and returns k ones that are most similar to the input. Similarity can be quantified using Euclidian or Manhattan distance. If the task is a classification problem, the algorithm returns the label most common within the k nearest neighbors. Similarly, If the task is regression, the algorithm returns their average.

5.3.2 Support Vector Machines

Support Vector Machines (SVM) are useful for classification problems. However, techniques used for support vector machines can also be used for regression problems, as support vector regression. In a space with N input dimensions, the SVM generates an N-1 dimensional hyperplane. For classification SVM, the hyperplane is optimized to have the most data points with one label on one side and the most data points with the other label on the other side. For regression, Least Squares Support Vector Machines (LSSVM) optimize the hyperplane to be as close as possible to all of the data points.

5.3.3 Neural Networks

The basic element of a neural network is an artificial neuron. As shown in Fig. 5.5, an artificial neuron has several inputs, denoted by x_1 to x_n. The inputs are each scaled by their respective weights, w_i, and then added to each other in a summing junction. The sum of the scaled inputs is then fed through a transfer function to give a single output, denoted by y.

A neural network consists of multiple instances of a neuron. In a neural network, the outputs of one neuron are fed as the inputs of another neuron. The most basic neural network, called a perceptron, consists of only a single neuron. There are different general structures of neural networks. Some relevant types will be discussed in Sects. 5.3.3.1 to 5.3.3.3.

Some of the most commonly used transfer functions in neural networks include the sigmoid function, the hyperbolic tangent (tanh), the rectified linear unit (ReLU), and the gaussian function, as outlined in Table 5.3.

The choice of an appropriate transfer function depends on many factors, including, but not limited to, the range of the desired output, the topology of the network, and the ease of training. In some design methodologies, an optimal transfer function can be found using trial and error.

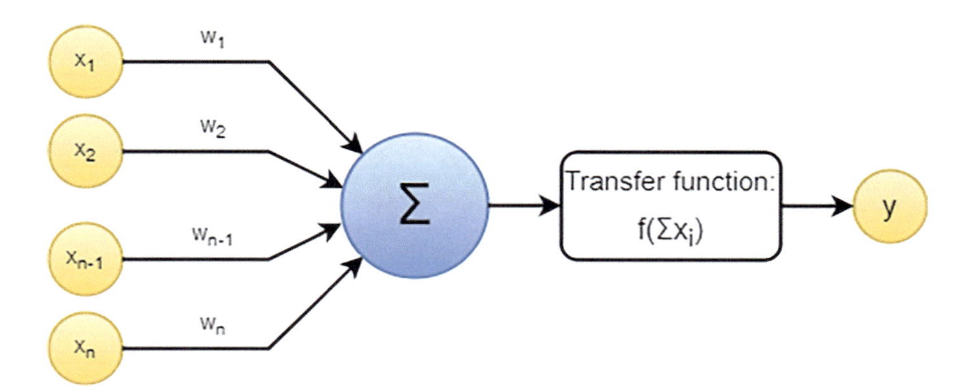

Fig. 5.5 Artificial neuron structure

Table 5.3 Common artificial neuron activation functions

Activation function	Sigmoid	Hyperbolic tangent	Rectified linear unit	Gaussian
Graph				
Equation	$S(x) = \frac{1}{1+e^{-x}}$	$\tanh(x) = \frac{e^x - e^{-x}}{e^x + e^{-x}}$	$\mathrm{ReLU}(x) = \begin{cases} x, x > 0 \\ 0, x \leq 0 \end{cases}$	$G(x) = e^{-x^2}$
Range	$(0, 1)$	$(-1, 1)$	$[0, \infty)$	$(0, 1]$
First derivative	$S'(x) = \frac{e^{-x}}{(1+e^{-x})^2}$	$\tan \mathrm{h}'(x) = 1 - \left(\frac{e^x - e^{-x}}{e^x + e^{-x}}\right)^2$	$\mathrm{ReLU}'(x) = = \begin{cases} 1, x > 0 \\ 0, x < 0 \end{cases}$	$G'(x) = -2xe^{-x^2}$

5.3.3.1 Feedforward Neural Networks

In a feedforward neural network (FFNN), information only flows in one direction. There exist no loops or feedback. Input nodes contain information fed into the network. Hidden nodes are neurons that receive information, perform computations, and send their output to other neurons for further processing. Output nodes receive information and perform computations just like hidden nodes. However, their output is visible to external users as the output of the network.

One important class of feedforward neural networks is the multilayer perceptron (MLP). As shown in Fig. 5.6, MLPs consist of nodes arranged in distinct layers. There exist a minimum of 3 layers, an input layer, a hidden layer, and an output layer. An MLP can have multiple hidden layers. If it has more than 3 hidden layers, the MLP is considered deep. Layers within an MLP are fully interconnected; the outputs of every node in a layer are visible only to every node within the next layer.

Organizing feedforward neural networks as multilayer perceptrons is very useful since, if the transfer function of all neurons within a layer is the same, it can take advantage of vector processing instructions within modern CPUs. One important subclass of multilayer perceptrons is radial basis function neural networks (RBFNNs). An RBFNN is limited to have only one single hidden layer, the radial basis layer, where the activation function is a radial basis function (RBF). An RBF is a symmetrical function whose output only depends on the distance of the input from the origin, such as the Gaussian function. The transfer function of the output nodes is linear, which means that the RBFNN outputs are linear combinations of the radial basis layer's outputs.

A derivative of the RBFNN is the general regression neural network (GRNN), first described by Specht [10]. It has a single hidden layer composed of neurons with

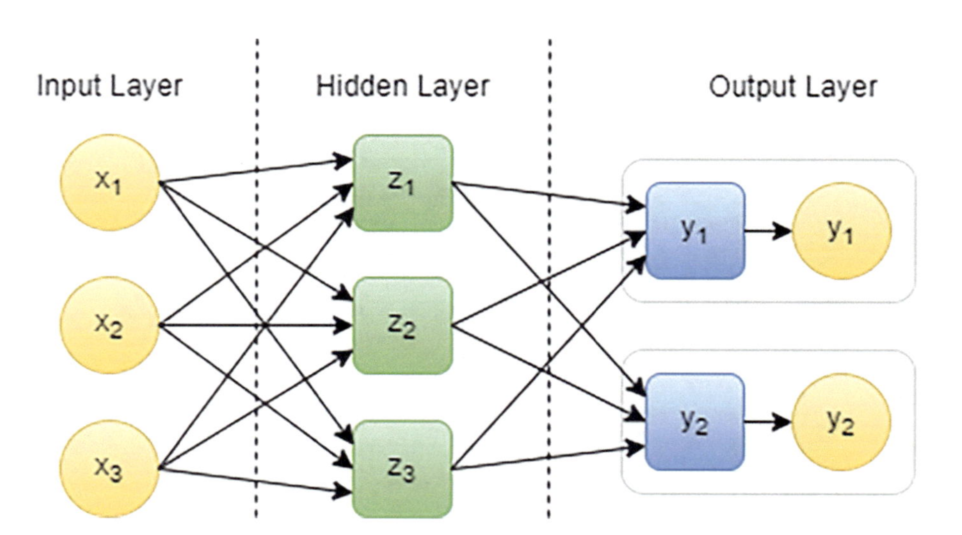

Fig. 5.6 Multilayer perceptron with 1 hidden layer

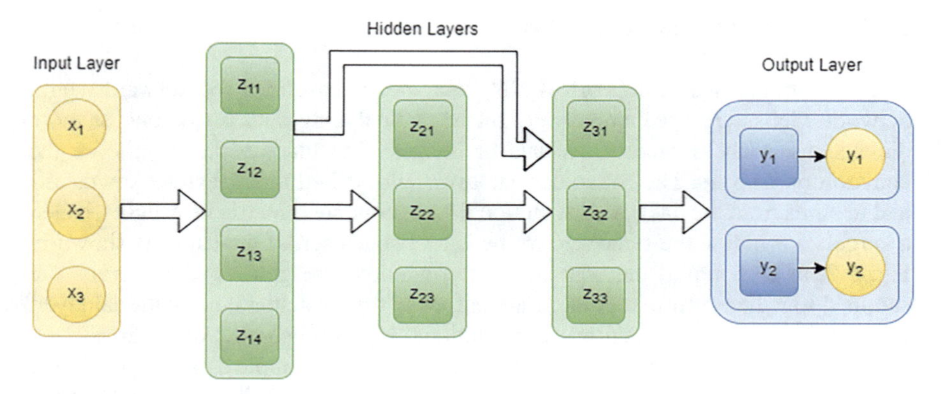

Fig. 5.7 Residual neural network

a Gaussian activation function. The difference between Gaussian RBFNNs and GRNNs is that while RBFNNs are trained using backpropagation, GRNNs are trained in one pass using probabilistic methods. Another important class of feedforward neural networks are residual neural networks (ResNN). Like MLPs, two layers that are connected to each other in ResNNs are fully interconnected. However, in ResNNs, layers are not limited to only being connected to adjacent layer in the series. Interconnections can skip layers, as can be seen in the example ResNN in Fig. 5.7.

Common methods of training feedforward neural networks are variations of gradient descent. However, for gradient descent, it is necessary to know the derivative of each individual weight and bias, with respect to the loss function. One common method of obtaining the parameter derivatives in feedforward neural networks is backpropagation. A neural network trained using backpropagation is called a back propagated neural network (BPNN).

In backpropagation, the chain rule is used to determine the derivative of the loss function with respect to every existing parameter. This vector containing these partial derivatives is the gradient. The dead neuron problem occurs when the derivative of a neuron's transfer function stays at zero when the system is given training data points. In this case, the neuron fails to update its weights and biases. A less severe variation to the dead neuron problem, vanishing gradient, occurs when the gradient of a function is very low. It can be trained, but the process is very slow. Networks built with one of the most common activation functions, ReLU, can suffer from dead neurons. To remedy this problem, variants that look similar to the ReLU function, but have consistently nonzero gradients, such as Leaky ReLU, Exponential Linear Unit [11], Gaussian Error Linear Unit [12], and Swish have been proposed [13]. Eliminating the vanishing gradient problem is a much arduous task, but Leaky ReLU or new oscillatory functions that have been proposed [14] can mitigate this problem.

5.3.3.2 Convolutional Neural Networks

A convolutional neural network (CNN) is a special case of a feedforward neural network. CNNs are exceptionally useful when analyzing data across one or more dimensions, such as time series data or images. The fundamental layers are the convolutional layers. The convolutional layers, also called filters, extract key details and features from the data. A convolutional layer contains a matrix of weights, called a kernel. A window the same size of the kernel slides across the data. As shown in Fig. 5.8, at each point, the contents of the window, weighted by the kernel, are summed together to form a point in the output of the particular convolutional layer. In most CNNs, there are multiple convolutional layers, in parallel or in series.

The next type of layer are pooling layers, which down sample data. The nonlinear layer adds nonlinearities to the data points. After processing by other parts of the CNN, the resulting data is fed into a fully connected neural network. Certain forms of data, such as chemical structures, are best represented as nodes interconnected to each other. There are different ways to represent these webs, including adjacency matrices and lists of interconnections. There is no unified standard structure for graph neural networks. Multiple architectures have been proposed for graph neural networks [15], including variants on convolutional neural networks [16] and recurrent neural networks.

5.3.3.3 Recurrent Neural Network

A recurrent neural network (RNN) is a neural network with feedback. An example is shown below in Fig. 5.9. During each time step, RNN neurons process their inputs into a value. At the end of the time step, they store this value, which then becomes available as an input for the same and other neurons. This is similar to how electronic flip flops store values in circuits when the clock signal changes.

In neural networks with feedback, the use of transfer functions that are not bounded on both sides, such as ReLU, is not recommended, since values can become arbitrarily large due to positive feedback, which causes instability.

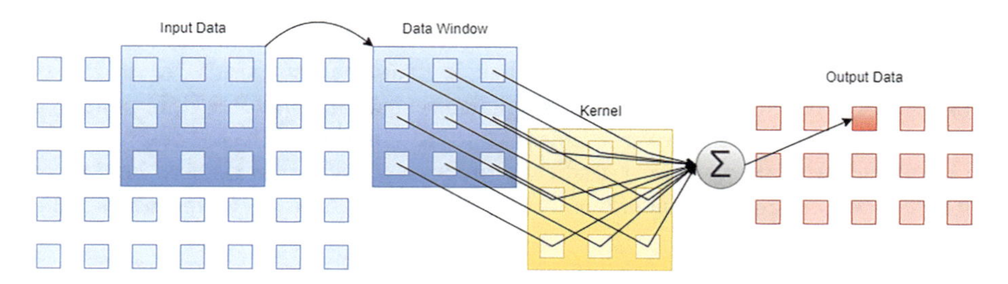

Fig. 5.8 Convolutional layer operation

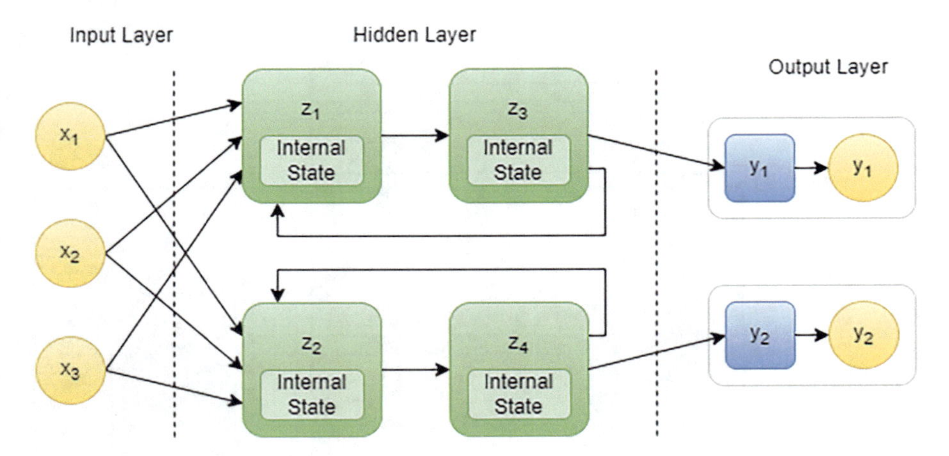

Fig. 5.9 Recurrent neural network

5.3.4 *Fuzzy Logic and Neuro-Fuzzy Systems*

In Boolean logic systems, variables or conditions can have one of two possible values, TRUE or FALSE. Boolean variables can be operated on by Boolean operators, such as NOT, AND, OR, and XOR. Fuzzy logic expands beyond these boundaries. The principle of fuzzy logic is that a condition may not be described discretely as TRUE or FALSE. It may be partially TRUE or partially FALSE. In fuzzy logic, variables are assigned values from 0 for completely FALSE to 1 for completely TRUE.

There are different ways of translating Boolean operators as fuzzy logic equivalents. One method is to interpret the fuzzy values as probabilities, the likelihood of a Boolean variable being TRUE. This way, the fuzzy logic operators can be inferred from the laws of probability. To translate numerical input values into fuzzy logic values, a process called fuzzification, input variables are operated on by multiple membership functions to obtain the values for each fuzzy variable. Common shapes for membership function are trapezoidal, triangular, and Gaussian. One ML model type that uses fuzzy logic is called the Adaptive Neuro-fuzzy Inference System (ANFIS). A diagram of the system is shown in Fig. 5.10. When training an ANFIS system, both the shapes of the membership functions and the rules may be changed.

5.3.5 *Decision Trees and Random Forest*

A decision tree is a type of machine learning model used for both regression and classification. An example of a decision tree, more specifically a binary decision tree, is shown in Fig. 5.11. Two types of nodes are present, decision nodes, indicated by brown diamonds, and leaves, indicated by green leaves. To use the decision tree to

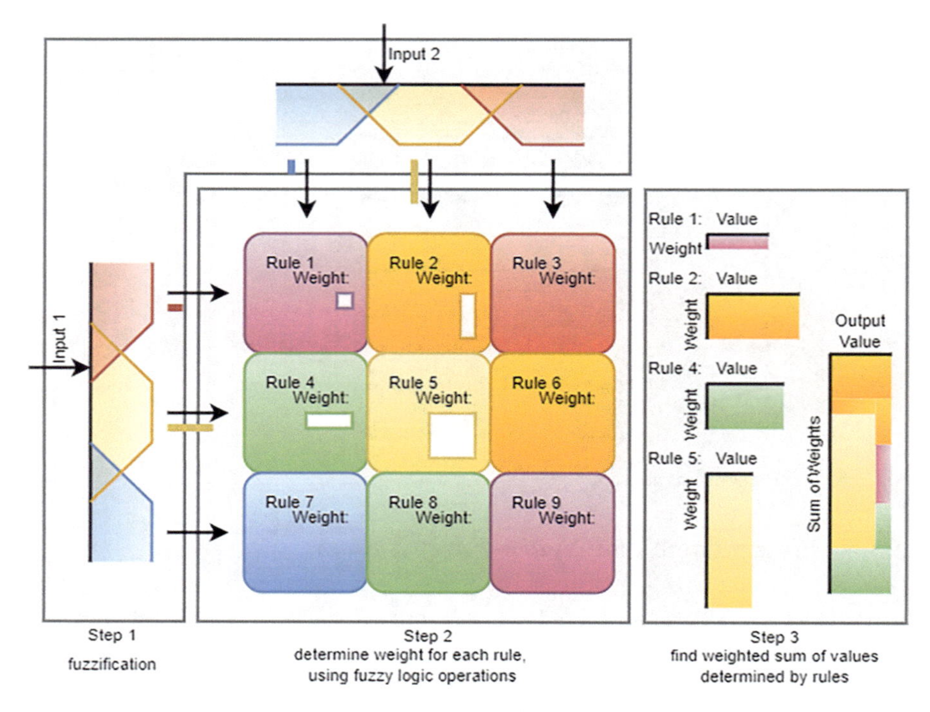

Fig. 5.10 Overview of the operation of an ANFIS system

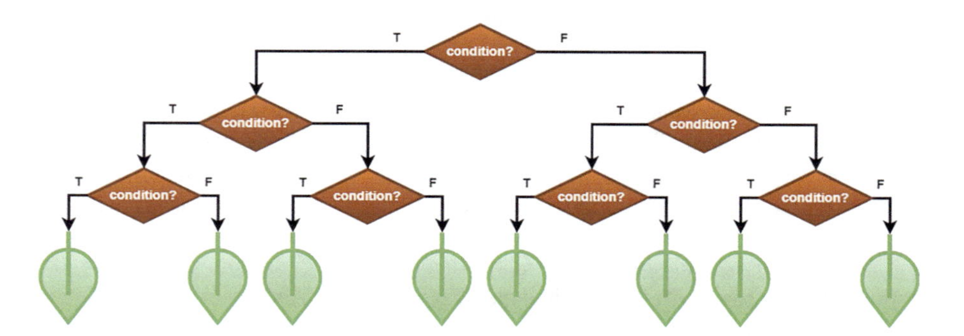

Fig. 5.11 Binary decision tree with a depth of 3

determine an output based on a set of inputs, start at the topmost decision node. Travel down a branch based on whether the condition stated by the node is true. If there is another decision node down the path, repeat the process. Eventually, there will be a leaf. This leaf contains the output of the model.

Training a decision tree involves the greedy algorithm. For each decision node, the algorithm computes the best split, the one with the lowest loss function, for each input variable. The variable with the split that has the lowest loss function then defines the decision node. The training algorithm then repeats this process for each side of the split, until it receives an instruction to not do so. Leaves are then placed in the sides of splits that do not split further.

Decision trees are prone to overfitting. To minimize the likelihood of overfitting, the tree depth can be limited. Additionally, each leaf can be required to represent a minimum number of training points. Random forests are a collection of trees trained in parallel on bootstrap aggregated training data. The output of a random forest is the average output of the ensemble of trees. Compared to decision trees, random forests are less prone to overfitting. An alternative to random forests is boosted trees. There are different tree boosting algorithms, including AdaBoost, Gradient Boosting, and XGBoost. In all boosting methods, weak learners, such as small decision trees, are utilized. Boosting is an ensemble learning method where the weight of every training point is adjusted after each round.

5.4 Prediction of Solvent and Solution Properties

One important step in the design of a carbon capture system is to determine which solutions will be used and to predict the properties of the solution. Factors which are vital in choosing an appropriate solution include mechanical properties, thermodynamic properties, solubility, kinetics, lipophilicity, and environmental considerations.

5.4.1 Mechanical Properties

Mechanical properties, such as density and viscosity, must be considered when choosing and configuring components that interact directly with the fluid. The solution needs to be pumped to the top of the absorber column. For a given volume, fluids that have higher densities have more mass and thus require more energy and power to be lifted. Similarly, a larger amount of force and energy is required to pump viscous fluids through a given gauge of pipe. Mechanical properties also affect reaction rates. Fluids that have higher viscosities tend to have droplets that are larger [17]. Larger droplets mean that the surface area to volume ratio is lower. This means that for a given volume of dripping fluid, the contact area it has with flue gas is lower. Since the CO_2 has to cross the surface before becoming absorbed by a droplet, a smaller surface area decreases the rate of absorption. Zhang et al. (2018) [18] trained both GRNN and BPNN neural networks to predict the density and viscosity of potassium lysinate and monoethanolamine solutions, based on solute concentrations and the temperature. Each neural network had at least four inputs: the concentration of potassium lysinate, the concentration of monoethanolamine, the sum of the concentrations of both solvents, and the temperature. The inclusion of the total solvent concentration as an input is redundant since the exact same information is available to neurons by increasing the weights of individual concentrations by the same value. The inclusion of this unnecessary input adds more weights and parameters to the model, increasing the computational demand of training.

Four datasets were used for the test and training data. Three datasets, which contained only potassium lysinate solutions, had diverse solvent concentrations, temperatures, densities, and viscosities. The one dataset that used the solutions of sodium lysinate mixed with monoethanolamine had relatively limited variation in the concentrations of either solvent, the temperature, the density, and the viscosity. This creates discontinuities in the regions of the input space where the model can be used. To avoid this problem, the model should have been focused on predicting the physical properties of solutions with only sodium lysinate solutions dissolved within them. For the density predictions, the GRNN network, with Gaussian activation functions, had the best performance. This model was accurate, with a small mean square error of less than 1% of every possible value and 2.7% of the range. In contrast, for the viscosity predictions, a single hidden layer BPNN network with 2 hidden neurons had the best performance. The author did not mention the activation function that was used in each layer. Overall, the models for the viscosity were far from precise. At best, they had a root mean square error of 0.24 ⍰/mPa s for typical values of 2 ⍰/mPa s. This means that neither the GRNN nor the BPNN with the given activation function was good models for viscosity.

5.4.2 CO_2 Solubility

CO_2 is soluble in water. However, its solubility is not high enough for feasible carbon capture purposes. For this reason, additives which increase the solubility of CO_2 in aqueous solutions are dissolved within water. Carbon dioxide reacts with primary and secondary amines to form carbamates and with tertiary amines and water to form bicarbonate. All of these species have higher solubility within water and thus serve as mediums for carbon dioxide transport in liquid flow. The amount of CO_2 that a given volume flow of liquid can transport in a given time frame depends on the solubility of CO_2 within that fluid. For this reason, it is important to predict the solubility of CO_2 within potential amine solutions. Hamzehie et al. [19] created a multiple layer feedforward network to calculate the equilibrium solubility of carbon dioxide, in mole ratios, within mixed solutions of 8 different amines and methanol. The network was trained using backpropagation, and the training and test data contained 505 points extracted from literature. The inputs consisted of temperature, CO_2 partial pressure, solute concentration, and apparent molecular weight, while the output was equilibrium CO_2 solubility. All input and output data points were linearly normalized on a scale from 0 for the minimum and 1 for the maximum. Multiple neural networks, with different numbers of neurons and hidden layers, and different transfer functions were trained and tested to determine the optimal configuration. They determined that the ideal network for this task contains 2 hidden layers, each consisting of 8 and 4 neurons with a tan sigmoid transfer function. This network was good at generalizing, with a 1.5% RMSE on the training set, and a 1.2% RMSE on the test set. However, caution should be used when utilizing this model, since only a limited number of solvent combinations were used in the training and test data. This model should not be expected to perform well when presented with a combination that it has not been trained with.

Similarly, Chen et al. [20] also used feed forward neural networks to predict the CO_2 solubility within solutions of 12 different amines. However, each solution only contained one solvent. Their study was more comprehensive, utilizing 2599 test and training points extracted from literature. Data points were also linearly normalized between -1 and 1. They developed and compared two different models for each solvent. The first model was trained using backpropagation. For each specific solvent, the ideal number of hidden layers, which was either 1 or 2, and the ideal number of hidden neurons, which ranged from 3 to 23, was unique. For the majority of solvents, the tansig function was the ideal transfer function. The exception was mono-diethanolamine, which yielded better performance with the logsig transfer function. The second model was a radial basis function neural network, with a Gaussian activation function. Similar to the BPNN, the ideal number of neurons in the RBFNN depended on the specific molecule. It ranged from 14 to 210. In both models, the ideal number of neurons was larger for molecules that had more data points extracted from literature. This discrepancy suggests that the neural network was performing simple regression, and not learning any fundamental rules about the chemical processes.

Li and Zhang [21] trained a generalized regression neural network to predict the CO_2 solubility in mixtures of trisodium phosphate and 6 amines. The inputs of the model included the concentrations of each of the 6 components, the total solute concentration, the temperature, and the CO_2 partial pressure. Train/test partitions of 70–30%, 80–20%, and 90–10% were used, with a resulting test set RMSE of 0.038, 0.035, and 0.032, respectively. The researchers concluded that the larger the training partition, the more accurate the model was, as indicated by the root mean square error of the test data. However, since the data set used for both training and testing was limited, consisting of only 299 points, this effect may be due to the number of training points being larger for larger training partitions. When training neural networks that are not too complex for the given problem, the learning process eventually reaches a plateau, where additional training points do not improve the performance. If the data set was larger, for the same train/test partitions, the number of training points would be larger. The network would possibly reach a training plateau and no more performance improvements could be made.

Yarveicy et al. [22] used four different machine learning models to predict the CO_2 solubility within piperazine solutions. These models included a least squares support vector machine (LSSVM), Adaboost regression trees, a neural network, and an ANFIS neuro-fuzzy system. The training and test data includes 577 points extracted from literature, with a wide range of temperatures, from 14 to 120 °C, pressures, from 0.03 to 9560 kPa, and solvent concentrations, from 0.1 to 7.7 mol/L. One potential bias is that studies that had lower solvent concentrations had lower partial pressures. This means that the training data is not homogenously distributed across the potential input space. An apparent correlation between the amine concentration and partial pressure can cause some models to associate effects due to one variable as being due to the other. Unfortunately, it is difficult to catch a degradation of performance due to this effect since the test also has the same bias. The authors fail to mention this potential bias, and present their data as being more diverse than it actually is. Three out of the four models did not show satisfactory performance. The average absolute relative deviance (AARD) for the LSSVM, neural network, and the neuro-fuzzy system were 16%, 19%, and 16%, respectively. In contrast, for the Adaboost CART model, the AARD was 0.9%. This is likely due to the fact that it was a consensus model, with 5 independently trained 14-layer decision trees. The biases that were present in individual trees likely cancelled out.

5.4.3 Reaction Rate

Solubility is not the only factor that is used to screen potential amine solution candidates. Fluids used in carbon capture are not stationary, allowed to sit until they reach equilibrium. They are dynamic and must move to the next stage of the process regardless of what fraction of their capacity has been used. If a hypothetical solution has a high solubility but slow kinetics, the main factor that limits how much CO_2 it can transport is the rate of dissolution. Machine learning can be used to identify novel structures which have faster kinetics than conventional solvents. Orlov et al. [23] used the physical experimental reaction rates of 24 conventional tertiary amine solvents and the in silico experimental reaction rates of 100 additional amines simulated in molecular dynamics software to train a random forest, a gradient boosted decision tree, and a support vector machine. The inputs were topological chemical structure descriptions, small substructures, and calculated physical and chemical properties. The outputs were the reaction rate and the free energy of absorption. Over 4000 tertiary amines were analyzed using the three regression tools to isolate potential candidates for further investigation. The reaction rate of the 18 candidate molecules were measured in physical experiments. Two amines, 1-methyl and 1-ethyl-3-pyrrolidinol, were found to have the fastest reaction rate.

5.4.4 Environmental Toxicity

The use of amines carries the risk of harming living organisms and, as a result, damaging the ecosystem. All organisms, including animals, plants, and bacteria, use amine containing organic molecules as internal signals to regulate their growth, development, and molecular processes necessary to keep them alive. Many organic amines bear resemblance to biological signaling molecules and, as a result, mimic their biological activity. This makes them capable of stimulating and inhibiting important biological signaling pathways. Chemists have synthesized amines which interact with these pathways to perform various tasks, which include killing unwanted organisms, such as invertebrates. An example of this is piperazine, a solvent currently in investigation for use in carbon capture, which agonizes the GABA receptor within invertebrates and has been used as an antihelminth drug [24].

Another mechanism that organic amines may harm living organisms is by causing damage to biological macromolecules. Some chemicals irreversibly bind to important biological molecules, such as enzymes, impairing their ability to perform important functions. Additionally, if a xenobiotic resembles a building block of macromolecules, it may become integrated within the backbone of a macromolecule. This can impair the function of the macromolecules or cause it to become prone to breaks. If a chemical increases the likelihood of certain segments of DNA becoming damaged or impairs the process of DNA repair, it will increase the probability that a mutant cell will replicate uncontrollably, inducing carcinogenesis. To minimize ecological effects from the usage of amine solvents, solvents should be chosen with the expectation that they will leak into environment. Amines should be screened for potential toxicity to ecologically significant species. Nontoxic or less toxic amines should be chosen for use. Machine learning can be used to screen novel solvents for potential toxicity. Liu et al. [25] developed multiple models to predict whether a chemical was toxic to crustaceans. These models included k-nearest neighbors, decision trees, random forest, naïve bayes, support vector machine, and neural networks. The inputs to the models were the chemical structure represented as a fingerprint and 10 different chemical descriptors based on topological, electronic, physical, and chemical properties. The output was a Boolean describing whether the chemical was acutely toxic to crustaceans at a concentration below 10 ppm. One strength of their model was that it could predict selective toxicity, when a chemical displays toxicity to only a subset of the studied species. Similarly, Xu et al. [26] developed multiple models, using the same six methods, to classify compounds based on their level of acute toxicity to honeybees. In this instance, acute toxicity was defined by the consumed amount of a substance that could be lethal to half of the population, the LD_{50}, within 48 hours. The input of the model was the chemical fingerprint, and the output was one of the following categories: nontoxic, with $LD_{50} \geq 11$ μg, moderately toxic, with $LD_{50} \geq 2$ μg, and highly toxic, with $LD_{50} \leq 2$ μg. In their study, they determined that the most reliable model for classifying acute toxicity was the support vector machine. One limitation of both models is that they only are only concerned with acute lethality. They do not consider whether chronic or repeated exposure to a chemical may have adverse effects on an organism's ability to survive and thrive. In addition, they do not provide insight into the potential mechanism of toxicity. The mechanism of toxicity is valuable information since if there are multiple toxins with similar, but not exactly the same, mechanisms, their effects can reinforce each other. Halabi et al. [27] used four different machine learning algorithms, based on decision trees, to predict if an aromatic amine is metabolized to a reactive species that can damage DNA and induce carcinogenesis. Their model not only indicates which amines have a carcinogenic metabolite, but it also identifies which of the specific metabolites is carcinogenic.

5.4.5 Lipophilicity

The water solubility and lipophilicity of a chemical determines whether it becomes distributed in water or soil. Species that live in free water are more vulnerable to hydrophilic toxins, while species that reside in soil are more vulnerable to hydrophobic toxins [25]. In addition, lipophilic species tend to become stored between fats. When they are consumed, organisms that were previously exposed to, and contain lipophilic toxins in their fatty tissues, pass the toxins on to the next trophic level. For this reason, fat-soluble toxins tend to bioaccumulate and reach greater concentrations in higher trophic levels. This makes carnivorous species more vulnerable to the harmful effects of lipophilic toxins. Information about solvent lipophilicity can also help make the carbon capture process more energy efficient. If a solvent in the lean phase is immiscible with the solvent in the loaded phase, the two phases can be mechanically separated. This can occur if the lean solvent is lipophilic and the loaded solvent is hydrophilic [28]. Only the hydrophilic fraction needs to be sent to the reboiler for regeneration, lowering the energy consumption of the reboiler [29]. Datta et al. [30] utilized both Mol2Vec and a deep multilayer perceptron to predict the lipophilicity of molecules, in terms of logP, the octanol-water partition. Mol2Vec is an unsupervised machine learning algorithm, based on language processing algorithms, which analyzed a vast database of molecules and assigned a 100 or 300-dimension vector identifier to each molecule based on its similarity with others [6]. The inputs of the MLP were the 300-dimensional Mol2Vec identifiers, and the output was the logP value. For the MLP neurons, the activation function that was used was ReLU. Different numbers of layers and neurons were tested.

The MLP with the best performance, the lowest RMSE of 0.627, had 10 hidden layers and 3225 hidden neurons. Since the logP is a logarithmic measure of lipophilicity, if two molecules have the same concentration in water and a logP difference of 0.627, the molecule with the higher logP will be 4.2 times more concentrated in octanol. This means that the RMSE was large. In addition, the training data consisted of 4200 molecules, while the fully connected 3225 neuron MLP included significantly more trainable parameters than the training parameters. This means that the neural network was likely overfitting to the training data. In addition, the authors noted that since the model uses the ReLU activation function, and it cannot have negative values, if the substance is hydrophilic, it predicts the logP to be 0. It seems that the transfer function that was chosen was not the most appropriate one. The authors should have experimented with different transfer functions to determine if they can find a network with better performance and less trainable parameters. Wieder et al. [31] trained multiple machine learning models to predict the lipophilicity and water solubility of organic chemicals. The inputs were either graphs of the chemical structures or chemical fingerprints combined with descriptors. The lipophilicity was stated as logD or logP, and the water solubility was stated as logS. Since these logarithms can have negative values, to make sure that every output value would be positive, the outputs were offset by the most negative value of each property. The training and test data included 4174 chemicals for the lipophilicity, and 6443 chemicals for the water solubility.

The models included 3 different types of graph neural networks (GNN), kNN, random forest, and support vector machines. Twelve to twenty-four instances of each type of network were trained, and the performance of the different types of networks was compared using RMSE. Both non-consensus and consensus models, which aggregate results from independently trained instances, were used. For all outputs, the most accurate model was consensus D-GIN, with an RMSE of 0.575, 0.455, and 0.738, for logD, logP, and logS. The second most accurate model was non-consensus D-GIN for logD, and consensus D-MPNN for logP and logS. In all cases, the most accurate non-consensus model, which was either the second or third accurate model overall, was D-GIN. Overall, the GNNs had significantly better performance than the other machine learning models. This may be due to the fact that the input variables were graphs for the GNNs, and fingerprints and descriptors for non-GNN models. The graphs may have contained relevant information that was not readily accessible with the fingerprints and descriptors.

5.5 Simulation and Analysis of Systems

Machine learning models trained on past data from existing carbon capture units can predict the performance, power requirements, and pollution release of the same and similar units. These models of carbon capture units can be used to optimize the performance of already existing units and to evaluate whether similar units should be produced.

5.5.1 *Multiple Predicted Variables*

Zhou et al. [32] created four different neural networks and neuro-fuzzy systems to predict the CO_2 production rate, the heat duty, and the absorption efficiency and the lean loading of a carbon capture system. They based their model off of 3 years of operational data from a pilot carbon capture plant. The inputs included pressures, CO_2 concentration, amine concentration, and flow rates. Originally, there were nine inputs, but two inputs were removed since they had negligible impact on the correlation coefficient of the neural networks. Removing extra inputs which do not provide additional relevant insight needed to make accurate predictions reduces the number of parameters in a given neural network. This reduces the chance that the neural network overfits to training data.

The neural network was very effective at predicting the carbon dioxide production rate of the system, with a correlation coefficient of 0.999. In contrast, it was less effective at predicting the lean loading, with a correlation coefficient of 0.884. This may be due to the presence of other unmeasured variables, which affect the lean loading. A secondary model, an ANFIS neuro-fuzzy model, was also created. Six variables were used as inputs into the system. Each variable was divided into three descriptive phrases, such as high, medium, and low. The membership in each description had a Gaussian distribution. The space of possible inputs was divided into 729 subspaces, each of which had their own rule. To train the model, a hybrid learning algorithm, consisting of back-propagated gradient descent and the least squares estimate, was utilized. Both the linear subspace rules and the exact shape of each membership distribution were adjusted. The neuro-fuzzy model had consistently high prediction accuracies and, with the exception of CO_2 production, was more accurate than the neural network model.

Machine learning can also be used to create models of components that can be integrated into other modeling software suites. Kim et al. [33] trained a deep residual neural network to model an absorber based on a pilot experiment with a novel water-lean diamine solvent, K_2Sol. The inputs of the model were the temperature of the flue gas, and properties of the lean amine solution, including mass flow, and the mole fractions of K_2Sol, water, and carbon dioxide. The outputs were the temperatures at different segments of the absorber, and the volume flow rate and CO_2 concentration of the vent gas. The neural network was trained and tested with 10,000 pilot experiment data points, using the Adam optimizer. The neural network absorber model was placed within a larger Aspen model. With the machine learning model, experiments were performed to explore the properties of the system. It was determined that, within the model based on an actual pilot system, the CO_2 absorption could not reach its full capacity. It was suggested that this means that the physical absorption tower should be made taller. Global sensitivity analysis was used to infer characteristics of the solvent, including vapor pressure and heat of formation, using the model. It was inferred that water and K_2Sol had similar absorption rates. One major oversight within this study was that the model was used in regions where experimental data was sparse or nonexistent, such as in areas with low liquid to gas ratios. In regions without training data points, a machine learning model, which is designed to perform regression, cannot provide meaningful results. Another issue with their methods was that the exact network structure, including the number of layers and the number of neurons in each layer, was specified before training. This may not be the optimal structure. If the method were to be improved, multiple structures would be trained in parallel, and their performance would be compared.

5.5.2 CO_2 Captured or Released

Li et al. [34] trained a single layer feedforward neural network using extreme learning machine techniques to obtain a carbon capture system model. Input variables included the reboiler duty, and both properties of the flue gas stream, such as temperature, pressure, flow rate, and CO_2 concentration, and properties of the amine stream, such as temperature, flow rate, and CO_2 loading. Outputs were the CO_2 production rate and the CO_2 capture rate. Multiple neural networks were trained in parallel using bootstrap aggregation. The output of the model was the average result of the neural networks. When testing with validation points, the aggregated networks had a consistently lower mean square error than each network alone. One advantage of their model was that it could accurately predict the dynamic CO_2 capture level up to 92 time-steps, or 460 s, ahead. Brauning et al. [35] used a long-short term memory network (LSTMN), a type of RNN, to predict the CO_2 concentration within the treated flue gas of an amine-based carbon capture unit. Time series data consisted of ten variables, including flow rates, pressures, and temperatures. The time steps were 20 min. After statistical analysis, only five of the variables, which had a significant effect on the output, were chosen to be used. This improves the model, since it reduces the number of parameters that have to be tuned, reducing the likelihood of the model overfitting. The ideal network, found using grid search, contained 6 hidden layers, 3 of which were LSTMN layers, and 3 of which were MLP layers. There were 448 LSTMN neurons and 896 MLP neurons, for a total of 1344 neurons. There was a total of 777,985 trainable parameters. To reduce the number of parameters that had to be trained, the network could have had symmetry where the values of certain weights are not independent. Overall, their model performed well for datapoints that were near the beginning of the time series, with relative concentration errors of 11%. However, one weakness of their method was that the RMSE error increased nearly linearly for samples that further along a time series. For a good recurrent neural network, error should not increase with time, it should reach a steady state amount.

5.5.3 *Power Requirements*

Moghadasi et al. [36] used historical data from a gas sweetening plant to create 10 different machine learning regression models that predict the steam consumption necessary for CO_2 removal. Discrete time values of variables were collected every 30 min for 4 years. The data was cleaned to remove noise and outliers. Hyper-parameter optimization was performed for each class of model using k-fold cross validation. Tree-based methods had significantly better performance than non-tree methods. The models that had the best performance, after hyper-parameter tuning, were random forest and gradient boosting method, which are both tree-based ensemble methods.

5.5.4 *Pollution Release*

Machine learning models can also be used to predict and estimate pollution resulting from a carbon capture system. Jablonka et al. [37] developed a convolutional neural network to predict amine emissions based on experimental measurements from a CO_2 capture pilot plant using the CESAR1 solution to process coal flue gas. This plant had the contents of its output lean flue gas measured by a continuous FTIR spectroscope. To organize the relevant data, they arranged plant process variables at a specific time into a vector. Process variable vectors for consecutive time stamps were aligned into a matrix. A convolutional neural network used for image recognition was modified, transformed from 2-dimensional to 1-dimensional, to analyze the matrix. Each time step represented 2 min. The matrix included parameters from the past 80 min and was used to predict the parameters for the next 4 min. It was trained on historical data. The network used to analyze the matrix had a depth of 8 layers, and each layer contained 128 filters. Each filter had a kernel size of 3 timestamps. Filters in the first layer analyzed consecutive time stamps, and filters in higher layers analyzed non-consecutive sequences of timestamps with holes between them. After the 8 layers, the results are aggregated into a single prediction. Separate CNNs were used to predict the amine, CO_2, and ammonia emissions of the plant 4 min, or 2 time-steps, into the future. One limitation of their model is that it assumes that the only way that the amines can leak into the environment is by evaporating into the flue gas stream. It assumes that the system that the amine loops through is perfectly sealed. A measure that would consider the total transfer of amines from the loop to the environment would be measurements of the volume and amine concentrations of the working solution.

5.6 Conclusion

In this chapter, a wide range of machine learning models in the field of post-combustion carbon capture were explored. In general, there were two types of models: static and dynamic. Static models were used to predict time invariant properties of both individual solvents and solutions, such as viscosity, equilibrium CO_2 solubility, and environmental toxicity. Dynamic models were used to predict time variable properties of both entire systems and individual components, such as CO_2 absorption rate and power consumption. Based off of common patterns between existing models, three recommendations can be made for creating future models. The first recommendation is to analyze the distribution of training and test data points. The most common oversight made while creating models was ignoring the domain of experimental data points. In most cases, data points were collected from different previously published experiments, but each experiment only covered a portion of the input space. There were parts of the input space that were not associated with any experimental data, where the models carry limited utility. In future studies, limitations due to the shape of the data distribution should be explicitly mentioned. The second recommendation is to set predefined boundaries for model complexity. The second common oversight was that models were overparametrized, leading to reductions in performance. Sometimes, hyperparameter optimization did not remedy this issue. Overparameterization can be avoided if there is a ceiling to the maximum number of adjustable parameters a model can have, lower than the number of training points. This way, the only way to improve the fit of a singular model would be to change characteristics which do not add additional trainable parameters, such as activation function, symmetry, and, in limited circumstances, topography. One pattern that stood out was that ensemble learning techniques consistently had better performance compared to their non-consensus counterparts. The third recommendation is to improve the performance of each type of model by training ensemble versions. Before creating a new machine learning model, it is good to note the strengths and weaknesses of previously created models. This way, the researcher can implement positive strategies that have been used in the past.

References

1. S. Solomon et al., Irreversible climate change due to carbon dioxide emissions. Proc. Natl. Acad. Sci. **106**(6), 1704–1709 (2009)
2. H. Ritchie and M. Roser, CO_2 and greenhouse gas emissions. *Our World in Data* (2020)
3. M.R. Smith, S.S. Myers, Impact of anthropogenic CO2 emissions on global human nutrition. Nat. Clim. Chang. **8**(9), 834–839 (2018)
4. I.C. Change, Mitigation of climate change. *Contribution of Working Group III to the Fifth Assessment Report of the Intergovernmental Panel on Climate Change* **1454**, 147 (2014)
5. R.R. Bottoms, Separating acidic gases. *Us1783901* (1930)
6. S. Jaeger, S. Fulle, S. Turk, Mol2vec: Unsupervised machine learning approach with chemical intuition. J. Chem. Inf. Model. **58**(1), 27–35 (2018)
7. D.P. Kingma and J. Ba, Adam: A method for stochastic optimization. *arXiv Preprint arXiv:1412.6980* (2014)
8. R. Ormiston et al., Noise reduction in gravitational-wave data via deep learning. Phys. Rev. Res. **2**(3), 033066 (2020)
9. Z. Zhao et al., Recurrent neural networks for atmospheric noise removal from InSAR time series with missing values. ISPRS J. Photogramm. Remote Sens. **180**, 227–237 (2021). https://doi.org/10.1016/j.isprsjprs.2021.08.009
10. D.F. Specht, A general regression neural network. IEEE Trans. Neural Netw. **2**(6), 568–576 (1991)
11. D. Pedamonti, Comparison of non-linear activation functions for deep neural networks on MNIST classification task. *arXiv Preprint arXiv:1804.02763* (2018)
12. D. Hendrycks and K. Gimpel, Gaussian error linear units (gelus). *arXiv Preprint arXiv:1606.08415* (2016)
13. P. Ramachandran, B. Zoph and Q.V. Le, Searching for activation functions. *arXiv Preprint arXiv:1710.05941* (2017)
14. M.M. Noel et al., Biologically inspired oscillating activation functions can bridge the performance gap between biological and artificial neurons. *arXiv Preprint arXiv:2111.04020* (2021)
15. J. Zhou et al., Graph neural networks: A review of methods and applications. AI Open **1**, 57–81 (2020)
16. T.N. Kipf and M. Welling, Semi-supervised classification with graph convolutional networks. *arXiv Preprint arXiv:1609.02907* (2016)
17. Z. Wang et al., Research on the effects of liquid viscosity on droplet size in vertical gas–liquid annular flows. Chem. Eng. Sci. **220**, 115621 (2020). https://doi.org/10.1016/j.ces.2020.115621
18. Z. Zhang et al., Machine learning predictive framework for CO_2 thermodynamic properties in solution. J. CO2 Utilizat. **26**, 152–159 (2018)
19. M.E. Hamzehie et al., Developing a feed forward multilayer neural network model for prediction of CO_2 solubility in blended aqueous amine solutions. J. Nat. Gas Sci. Eng. **21**, 19–25 (2014)

20. G. Chen et al., Artificial neural network models for the prediction of CO_2 solubility in aqueous amine solutions. Int. J. Greenh. Gas Control **39**, 174–184 (2015)
21. H. Li, Z. Zhang, Mining the intrinsic trends of CO_2 solubility in blended solutions. J. CO2 Utilizat. **26**, 496–502 (2018)
22. H. Yarveicy, M.M. Ghiasi, A.H. Mohammadi, Performance evaluation of the machine learning approaches in modeling of CO_2 equilibrium absorption in Piperazine aqueous solution. J. Mol. Liq. **255**, 375–383 (2018)
23. A.A. Orlov et al., Computational screening methodology identifies effective solvents for CO_2 capture. Commun. Chem. **5**(1), 1–7 (2022)
24. R.J. Martin, Modes of action of anthelmintic drugs. Vet. J. **154**(1), 11–34 (1997)
25. L. Liu et al., In silico prediction of chemical aquatic toxicity for marine crustaceans via machine learning. Toxicol. Res. **8**(3), 341–352 (2019)
26. X. Xu et al., In silico prediction of chemical acute contact toxicity on honey bees via machine learning methods. Toxicol. In Vitro **72**, 105089 (2021)
27. A. Halabi, E. Rincón, E. Chamorro, Machine learning predictive classification models for the carcinogenic activity of activated metabolites derived from aromatic amines and nitroaromatics. Toxicol. In Vitro, 105347 (2022)
28. Y. Coulier et al., Thermodynamic modeling and experimental study of CO_2 dissolution in new absorbents for post-combustion CO_2 capture processes. ACS Sustain. Chem. Eng. **6**(1), 918–926 (2018)
29. L. Raynal et al., The DMX™ process: An original solution for lowering the cost of post-combustion carbon capture. Energy Procedia **4**, 779–786 (2011)
30. R. Datta, D. Das, S. Das, Efficient lipophilicity prediction of molecules employing deep-learning models. Chemometr. Intell. Lab. Syst. **213**, 104309 (2021)
31. O. Wieder et al., Improved lipophilicity and aqueous solubility prediction with composite graph neural networks. Molecules **26**(20), 6185 (2021)
32. Q. Zhou et al., Modeling of the carbon dioxide capture process system using machine intelligence approaches. Eng. Appl. Artif. Intell. **24**(4), 673–685 (2011)
33. J. Kim et al., Learning the properties of a water-lean amine solvent from carbon capture pilot experiments. Appl. Energy **283**, 116213 (2021)
34. F. Li et al., Modelling of a post-combustion CO2 capture process using extreme learning machine. Int. J. Coal Sci. Technol. **4**(1), 33–40 (2017)
35. L.F.G. Brauning et al., Application of long short-term memory neural networks for CO_2 concentration forecast on amine plants
36. M. Moghadasi, H.A. Ozgoli, F. Farhani, Steam consumption prediction of a gas sweetening process with methyldiethanolamine solvent using machine learning approaches. Int. J. Energy Res. **45**(1), 879–893 (2021)
37. K.M. Jablonka et al., Deep learning for industrial processes: Forecasting amine emissions from a carbon capture plant (2021)

Chapter 6
Design and Optimization of a Tidal Power Generation Plant in the Bay of Fundy, Canada

Reagan McKinney, Claudia Nashmi, Arash Rafat, Ali Ahmadian, and Ali Elkamel

Nomenclature

Variable description	Symbol	Units
Area of the turbine swept range	A	[m^2]
Area of the channel	A_c	[m^2]
Blockage ratio	B	[−]
Capital cost of the turbine	$C_{turbine}$	[$]
Density of seawater	ρ	[kg/m^3]
Depth of the turbine	d	[m]
Diameter of the turbine	D	[m]
Fraction of the turbine operation yearly	f_y	[−]
Gravitational constant	g	[m/s^2]
Generator efficiency	η	[−]
Power generated	P	[kW]

(continued)

R. McKinney · A. Rafat
Department of Civil and Environmental Engineering, University of Waterloo, Waterloo, ON, Canada
e-mail: rmckinney@uwaterloo.ca

C. Nashmi
Department of Chemical Engineering, University of Waterloo, Waterloo, ON, Canada

A. Ahmadian (✉)
Department of Electrical Engineering, University of Bonab, Bonab, Iran

Department of Chemical Engineering, University of Waterloo, Waterloo, ON, Canada

A. Elkamel
Department of Chemical Engineering, University of Waterloo, Waterloo, ON, Canada

Department of Chemical Engineering, Khalifa University, Abu Dhabi, United Arab Emirates
e-mail: aelkamel@uwaterloo.ca

A. Ahmadian et al. (eds.), *Carbon Capture, Utilization, and Storage Technologies*, Green Energy and Technology, https://doi.org/10.1007/978-3-031-46590-1_6

Variable description	Symbol	Units
Return period	R	[years]
Revenue of energy sold	C_{energy}	[$]
Theoretical turbine power	P_{ext}	[kW]
Thrust coefficient	C_T	[−]
Tip speed ratio (TSR)	λ	[−]
Turbine power coefficient	C_p	[−]
Turbine thrust	T	[kg·m/s^2]
Turbine wake	u_{4t}	[m/s]
Wake speed ratio	α_4	[−]
Water flow velocity	u	[m/s]
Water velocity through the turbine	u_{2t}	[m/s]

6.1 Introduction

In today's society, Canada is estimated to produce roughly 2% of global electricity, about 641.1 (TWh) of electricity [24]. Of this produced electricity, Natural Resources Canada [24, 25] estimates that roughly 67% of electricity produced in 2018 was from renewable energy sources, including hydro power (non-tidal) and other non-hydro power means of electricity generation. Currently, Canada's renewable electricity production has largely focused on wind, biomass, and solar (photovoltaic). As of 2018, only 0.1% of total renewable energy comes from tidal energy generation [25]. With the world's largest coastline spanning three oceans [35], Canada has a large potential to develop the tidal energy and electricity generation industry. With a desire for the Canadian government to drastically reduce emissions to meet the Paris Climate Agreement, the expansion of the tidal power generation capacity could provide a sustainable means of meeting future electricity demands.

Tidal power generation involves the conversion energy within tides to useful electricity. Two general systems are applied for tidal power generation: barrages and tidal stream systems [16]. Barrages use the potential energy from differences in hydraulic head across a structure associated with low and high tides. Although effective, these systems have high associated infrastructure costs [16] and potential environmental effects such as alteration of marine habitats and interactions of wildlife with the turbine rotors [27]. On the other hand, tidal streams focus on harnessing the kinetic energy of the tides. Similar to wind turbines, these systems use water currents to move a turbine to generate electricity and have lower environmental impacts and lower costs than the barrage-type power plant design [16]. These systems have been rapidly developing and have been the focus of many, as these systems are less constrained than their barrage-typed design counterparts. Additionally, difficulties arise for barrage systems as the geography of a region dictates the viability of a barrage-type tidal power generation system which typically requires tidal ranges to be sufficiently large to produce power in an economically viable manner.

Presently, there exist only 5 large-scale tidal power plants around the world, the largest of which is the Shiwa Lake plant in South Korea with a 254 MW capacity and a tidal barrage design [3]. The third largest tidal generation plant in the world is the Annapolis Royal Generating Station along the Bay of Fundy. This system is an ebb-tide barrage-type operation with a basin area of 6 km^2 basin area and a generation capacity of 20 MW. The barrage type design may not be necessary for small-scale community-based generation of tidal energy. As a more promising solution for smaller-scale energy generation needs, the usage of tidal-stream-based turbines have been shown to be effective [33]. An example of this would be the Strangford Lough in Northern Ireland where the SeaGen device has been successfully installed [6, 33]. These devices have much greater predictability as the systems only rely on current velocities rather than tidal ranges. For this technology, however, economic feasibility remains a challenge as the cost of tidal-stream energy generation is highly dependent on external factors such as flow velocity distributions, direction of flow, water depth, construction and capital costs, geotechnical properties, and the distance from shore.

This study aims to design and optimize a small-scale tidal-stream energy generation facility for a small community ($\leq$5000 population) along the Bay of Fundy in Nova Scotia. The Bay of Fundy has one of the largest tidal ranges in the world [38]. In addition, the unique geometry of the Bay of Fundy makes it an excellent candidate for the design of not only barrage-type tidal energy generation systems which rely on large tidal ranges but also tidal-stream-type systems which would benefit from large tidal-stream velocities. A non-linear programming (NLP) model was developed to maximize energy generation while meeting a series of societal, design, and economic constraints. The results of this study will help elucidate the viability of tidal-stream energy generation facilities as a means of achieving sustainable energy for small communities. Further, these results will assist policy makers and planners with Canada's transition away from fossil fuel–based energy production.

The chosen location for the turbine is the Minas Passage in the Bay of Fundy. More specifically, the turbine will be placed on the south-eastern portion of the channel, just off the coast of Scot's Bay. This location was chosen due to the depth of the water, the speed of the tidal activity, and the proximity to nearby communities [7, 14, 15]. This portion of the passage is quite narrow and therefore transmission lines can be sent to either the west or east coasts to accommodate more communities.

The turbine design in this study will be based on the usage of a crossflow turbine designed to produce enough energy to meet the demands of a community of approximately 5000 people, typical for a small community close to the chosen site in the Bay of Fundy. According to data from the Canada Energy Regulator, the annual per capita consumption for the residential sector in Nova Scotia was 4.5 TWh of electricity in 2017 [5]. Based on this estimate, the minimum required capacity of the turbine is 2.7 MW.

Water speeds in the channel vary over depth and over the daily cycle due to tidal activity. During ebb flow (low-tide), which lasts for 4 hours per day, the water speed is 1–1.5 m/s slower compared to flood flow (high-tide) [31]. The variations in water

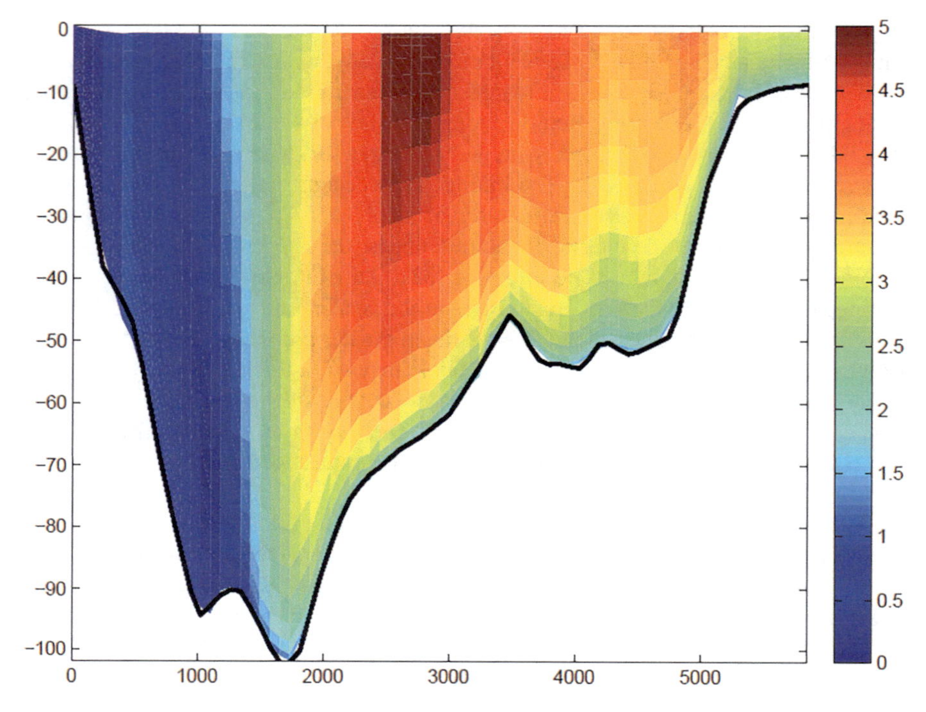

Fig. 6.1 Depth distribution of water speeds in the Minas Passage [14, 15]

speed during high tide are shown in Fig. 6.1. Depth distribution of water speeds in the Minas Passage [14, 15], with the higher speeds occurring near the top of the water column due to reduced friction from the bottom of the passage [14, 15].

When considering transportation needs, the Norwegian Government regulates that all ocean infrastructure constructed with steel be placed at a minimum depth of 18 m below the water surface to accommodate shipping routes and other watercraft [8]. The conference on Contemporary Achievements in Civil Engineering further states that a depth of 20 m below the surface is more optimal [30]. No regulations with respect to the required depth for the placement of ocean infrastructure to allow for safe transport via water were found. Further, it should be noted that the turbine will require its transmission lines and anchor to be placed on the ocean floor, so it would be cost-effective to have it situated as close to the bottom as possible. However, the shear stress of the ocean floor will need to be examined as well in order to ensure that the drag of the boundary layer will not interfere with optimum power generation from the turbine.

Extensive research has been conducted on the cross-axial tidal turbine and its parameters. Unlike wind turbines which are highly susceptible to drag forces, tidal turbines prefer higher blockage ratios because it increases the flow velocity through the turbine [39]. Thus, the recommended design will include the highest blockage ratio available given the geometry of the channel, waterway regulations, and cost of the turbine. For design considerations, the tip speed ratio, or the ratio of the angular velocity of the tips of the turbine to the tidal velocities, was selected to be between 4 and 6. However, tip speed ratios greater than 6 are very susceptible to cavitation [42].

Lastly, it is imperative that total costs associated with the project (capital and operations and maintenance costs) are minimized in order to ensure that the power plant will be feasible for the small surrounding communities. The current levelized capital cost for a cross-flow turbine is \$0.48/kWh, which is equivalent to \$1150/kW of the system [19]. In contrast the average price of electricity per kWh in Nova Scotia is between \$0.08 and \$0.20 depending on when the energy is being used (peak vs. non-peak hours) [26]. Therefore, a deficit is expected during the first few years of the turbine's life regardless of optimization of the turbine design.

6.2 Methodology

6.2.1 Turbine Design

For this study, a crossflow turbine was designed for deployment in the Minas Channel, located in the Bay of Fundy, Canada. The turbine was designed according to Betz Theory [4] for the theoretical limit for an isolated wind turbine's efficiency, as wind turbines and tidal turbines operate on the same general principles. This theory considers steady-state flow in an adequately large and deep pool such that bottom and side effects of walls are not experienced by the flow passing through the turbine [14, 15]. A derivation of Betz Theory by Kartsen et al. [14, 15] is summarized hereafter. Flow dynamics are derived through the conservation of mass, momentum, and energy (Bernoulli's Equation). The energy that a turbine can take from the flow is given by Eq. (6.1) below, where α_4 represents the power extraction coefficient, u_{t4} represents the water speed in the turbine wake [m/s], and u is the upstream water speed [m/s].

$$\alpha_4 = \frac{u_{t4}}{u} \tag{6.1}$$

The flow that is going through the turbine (u_{t2})[m/s] is therefore given by Eq. (6.2).

$$u_{t2} = \frac{u}{2}(1 + \alpha_4) \tag{6.2}$$

Turbine thrust (T) [N], or the force of the water on the turbine, is calculated by Eqs. (6.3), where A represents the swept area of the turbine blades [m^2] (Eq. 6.4), ρ is the density of seawater [1025 kg/m^3], and C_T [−] is the turbine thrust coefficient and can be calculated using Eq. (6.5).

$$T = C_T\left(\frac{1}{2}\rho A u^2\right) \tag{6.3}$$

$$A = \frac{\pi D^2}{4} \tag{6.4}$$

$$C_T = 1 - \alpha_4^2 \tag{6.5}$$

The power generated (P) by an individual turbine is given as the product of the turbine thrust (T) and the flow through the turbine (u_{t2}) resulting in the simplified Eq. (6.6). Cp [−] represents the power coefficient and is given by Eq. (6.7).

$$P = C_P \frac{1}{2}\rho A u^3 \tag{6.6}$$

$$C_P = \frac{C_T}{2}(1 + \alpha_4) \tag{6.7}$$

The power generated by the turbine has a theoretical upper limit as turbines are only able to use a proportion of the available energy. This theoretical upper limit is referred to as the extracted power, or P_{Ext}, and is calculated through Eq. (6.8).

$$P_{Ext} = C_T \frac{1}{2}\rho A u^3 \tag{6.8}$$

This is typically the case due to energy losses due the mixing in the wake region downstream of the turbine resulting in a lower proportion of power remaining for use [14, 15]. Through Eqs. (6.4) and (6.6), it is evident that (C_P) can only take a maximum value of $\frac{16}{27}$ or approximately 0.59. This is referred to as the Betz limit and is the theoretical upper bound on the conversion of kinetic energy of a moving fluid to electrical energy through a turbine. The efficiency of the turbine is the relationship between P_{Ext} and P, as given below in Eq. (6.9).

$$\eta = \frac{P}{P_{Ext}} \tag{6.9}$$

where η is the efficiency of the turbine [−]. It should be noted that the drag coefficient plays an important role in determining the power generation potential of an isolated turbine. This influence is noted by choice in the shape and number of blades chosen for the design. Kurniawati et al. [18] show that using a 16-blade crossflow turbine design shows optimal performance in terms of maximizing C_P and power generation. Therefore, for this design, a 16-blade cross flow design will be considered.

6.2.2 *Optimization*

In order to design an optimum solution to meet the needs of the small communities in Nova Scotia, an NLP model was developed. The objective function is set up such that the profits gained from the turbine are maximized and the total capital and O&M costs over the desired design life of the turbine are minimized. The objective function is presented below in Eq. (6.10).

$$Z = C_{energy} - C_{turbine} \tag{6.10}$$

where C_{energy} is the revenue received from selling the energy produced from the turbine [$] and $C_{turbine}$ is the total cost of the turbine [$]. The capital cost of the turbine was estimated to be $1500 per kilowatt of rated power [19] and the estimated revenue is between $0.08/kWh and $0.20/kwh of energy produced, as per the current rates of energy in the Province of Nova Scotia [26]. As a result, the objective function can further be broken down into the following Eq. (6.11).

$$Z = C_{power} \times 8760 \times P - 1500 \times P_{ext} - OM \tag{6.11}$$

where P is the actual power generated from the system [kW], P_{ext} is the rated power of the system in an ideal scenario [kW], C_{power} is the revenue associated with selling power to the grid, and OM is the O&M costs associated with the project. Further parameters were also included in the model such as an operation fraction to account for the portion of the year that the turbine would be running.

6.2.2.1 Constraints

This objective function is subject to various constraints as outlined in the following section. The speed of the water is highly related to depth, so in creating the objective function constraints, an empirical relationship between the depth of the water and its flow velocity was established with an R^2 value of 0.97. The relationship was adapted from [14, 15] and is presented below in Eq. (6.12).

$$u = 0.0475d + 5.4861 \tag{6.12}$$

where u is the velocity of the water and d [m] is the depth the turbine is situated at below the surface of the water. Similarly, the diameter of the turbine blades was determined through an empirical relationship between the depth of the turbine in order to ensure the blades would not protrude the surface of the water, nor reach the ocean floor. The relationship is presented in Eq. (6.13) below and has an R^2 value of 0.93.

$$D = -0.0562d^2 - 6.6571d - 131.24 \tag{6.13}$$

where D [m] is the diameter of the blades. This equation along with the water velocity empirical relationship is subject to heavy constraints though to account for the structure being bounded by the ocean surface and ocean floor. The depth (d) of the turbine beneath the ocean surface must be greater than 20 m to accommodate best practices for offshore infrastructure [8], but less than 98 m below the surface to ensure that the blades do not interfere with the ocean floor, which is 100 m, below the surface. The power equation also considers various coefficients which must be between 0 and 1 in order to be effective in the calculations (α_4, C_T, T, η, C_P, α_2, B). Further, the upstream water velocity (u) must be greater than the flow through the turbine (ut2), and the flow through the turbine must be greater than the turbine wake (u_{t4}). Lastly, the power generated by the turbine must be greater than 2.7 MW in order to satisfy the energy needs for the scope of this work. The following Eqs. (6.14, 6.15, 6.16, 6.17, 6.18, 6.19, 6.20, 6.21, 6.22, 6.23, 6.24, 6.25, 6.26, 6.27 and 6.28) written as inequalities represent these constraints below.

$$d < 98 \tag{6.14}$$

$$d > 20 \tag{6.15}$$

$$u > ut2 \tag{6.16}$$

$$ut2 > ut4 \tag{6.17}$$

$$P > 2700 \tag{6.18}$$

$$\alpha 4 > 0 \tag{6.19}$$

$$\alpha 4 < 1 \tag{6.20}$$

$$CT > 0 \tag{6.21}$$

$$CT < 1 \tag{6.22}$$

$$\eta < 1 \tag{6.23}$$

$$\eta > 0 \tag{6.24}$$

$$B > 0 \tag{6.25}$$

$$B < 1 \tag{6.26}$$

$$CP > 0 \tag{6.27}$$

$$CP < 1 \tag{6.28}$$

It should be noted that blockage ratio (B) was not used in the power calculations but was utilized in the constraints to ensure that the design fits within the specified channel as Betz theory assumes a blockage ratio of zero. Definitions for all of the

Table 6.1 Summary of turbine design

Design parameter	Value
Turbine diameter (D)	45 m
Placement depth (d)	75 m
Swept turbine area (A)	1623 m^2
Flow through turbine (u_{2t})	1.28 m/s
Upstream water velocity (u)	1.92 m/s
Power coefficient (C_P)	0.59
Power generated (extracted power) (P)	3.5 MW
Rated power (P_{Ext})	5.3 MW

parameters may be found in Table 6.1 of this report. The optimization of the design was then carried out using Excel.

6.2.2.2 Generalized Reduced Gradient Method

To solve the developed optimization problem at hand, the Generalized Reduced Gradient Method (GRG) was adopted and used in Excel. This solution method is a non-linear solving algorithm that considers the gradient of the objective function in order to find locally optimal solutions [32]. As a basic overview, the GRG method operates by separating decision variables into basic and non-basic sets, and a reduced gradient computed to determine the direction of travel to the local minimum or maximum [23]. To ensure that a feasible solution for this optimization problem is near the optimal solutions, the GRG was repeated for changing initial conditions until a series of solutions were obtained. The feasible solutions were then compared to each other and the solution that best fulfills the objectives of this study was selected.

6.2.3 Projected Lifetime Costs and Payback

In order to determine the true feasibility of the turbine, the return period and net present value of the turbine were also taken into consideration. Operations and maintenance costs for a small turbine to fill these goals were collected from Li and Florig [20, 21] who determined the empirical relationship between operations and maintenance costs over the lifetime of the turbine. The return period was estimated using Eq. (6.29) and included both the capital and operations and maintenance costs of the turbine.

$$R = \frac{C_{turbine}}{C_{energy}} \tag{6.29}$$

where R is the return period [years], $C_{turbine}$ is the cost of the turbine [$], and C_{energy} is the revenue received from the energy that turbine produces [$]. Net present value is associated with Eq. (6.30) below.

$$NPV = \frac{R_t}{(1+i)^t} \tag{6.30}$$

where NPV is the net present value of the turbine [$], R_t is the current value of the turbine [$] including profits, i is the discount value [%], and t is the time period being examined for the turbine. The current best practices for turbine feasibility studies suggest a discount rate of 10% [2, 17, 22], which was also adopted for this analysis. A time period of 20 years was also utilized following the common lifetimes of tidal turbines. Operations and maintenance costs were estimated to be 0.05¢/kwh produced for a small turbine farm [20, 21]. These costs were discounted using a 10% interest rate and converted to a present value. The net profits of the project were estimated by subtracting the net discount revenue over a 20-year design life minus the capital and discounted operations and maintenance costs. It should be noted that a constant discount rate was assumed across the entire design life of the project and the influence of inflation and taxes was not considered. Additionally, a constant price per electricity produced from the tidal turbine was adopted for the operations and maintenance costs.

6.3 Simulation Results

6.3.1 Model Design and Optimization

Results from this study indicate that a tidal turbine placed 75 m below the surface of the water with a diameter of 45 m would produce enough electricity to power a small community of approximately 5000 people. The turbine is estimated to produce 3.5 MW of power, thereby exceeding the minimum power requirement of 2.7 MW. A summary of turbine design variables and properties of interest are summarized in Table 6.1.

It should be noted that the turbine was designed such that the generated power (P) was greater than the minimum 2.7 MW requirement to meet the needs of a 5000 people community and not the extracted power (P_{Ext}). To achieve this result, the maximum extracted power (P_{Ext}) was constrained to a maximum power generation capacity of 3.6 MW. This constraint was applied to determine the minimum requirements for meeting the anticipated demand of 2.7 MW. In practice, the turbine can be up-scaled to produce a much greater power generating capacity. Figure 6.2 shows the power generating capacity for different turbine diameters and placement depths in the water column.

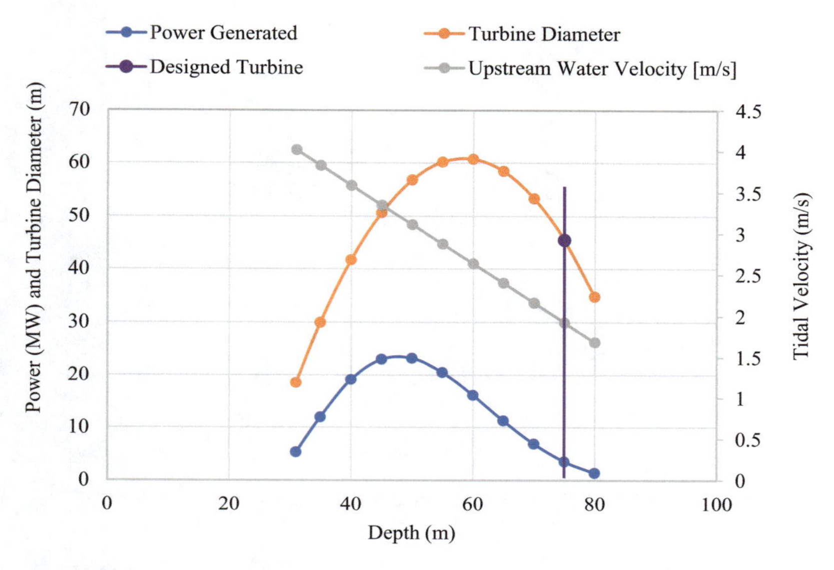

Fig. 6.2 Turbine design for varying depths

From Fig. 6.2, it is noted that both the turbine diameter and power generated from the turbine follow polynomial functions as a function of depth. Peak power generation, however, does not correspond with maximum turbine diameters, nor with the maximum tidal velocities. This design can produce a minimum of 1.3 MW with a turbine diameter of approximately 35 m with the center the of turbine placed 80 m below the surface water level. The maximum possible power generation capacity using a single cross flow turbine is 23.2 MW and would require a 56.8 m diameter turbine blade and an upstream tidal velocity of approximately 3.1 m/s. Tidal velocities were estimated to be between 1.7 and 4.0 m/s and linearly decrease with depth with maximum and minimum velocities occurring at the depths of 31 m and 80 m, respectively. The selected optimal turbine design for this project is shown via the purple line in Fig. 6.2.

6.3.2 *Economic Analysis*

Results from the optimization of the turbine design and the economic analysis suggest that a total profit of $44.7 million is expected when considering a 20-year design life. This design, through the conducted economic analysis, has a return period of approximately 1.28 years. This profit consists of the difference between the present value of the capital cost of $7.9 million, annual profits from energy sold to the Nova Scotia grid ($6.2 million), and annual operations and maintenance cost of approximately $6300. A cost of $0.20/kwh was used to represent the cost of energy sold to the Nova Scotia grid. Figure 6.3 provides a visual representation of the net

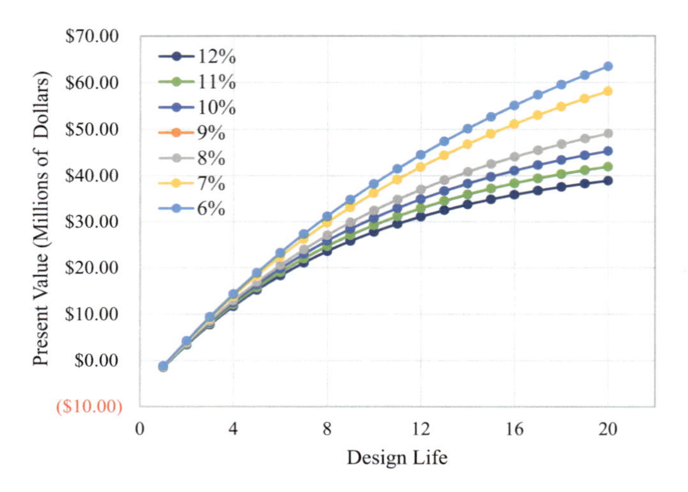

Fig. 6.3 Net present value – varying discount rates and design lives

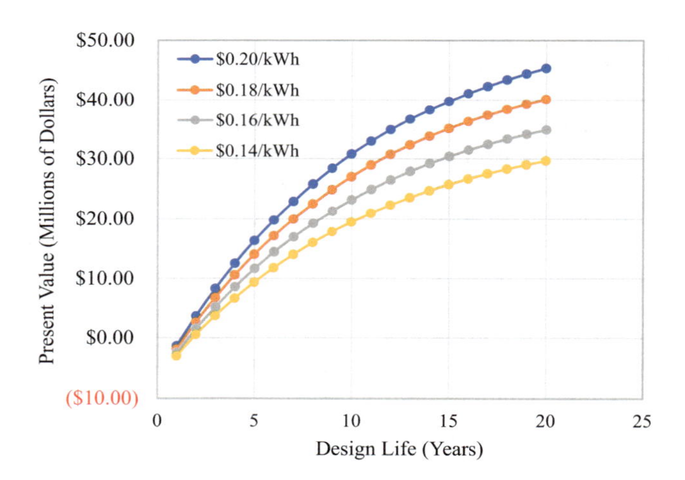

Fig. 6.4 Net present value – impact of shifting grid prices

present value as a function of design life. A sensitivity analysis was conducted to see the impact of total costs and profit for varying discount rates.

From Fig. 6.3, it can be noted that the present value of the project increases as the design life of the project is extended. Similarly, a decrease in discount rate increases the present value. For example, net present values of the project for a 6% discount rate versus a 12% discount rate yields a difference of approximately $25 million dollars for a 20-year design life. On the other hand, when considering only a 5-year design life, this difference drops to only approximately $3.7 million. To determine the impact of shifting electricity selling prices on the net present value of the project, a sensitivity analysis was conducted. Figure 6.4 provides the results of this analysis in which the cost of selling electricity to the grid was reduced by increments of $0.02/kWh until the current 2020 standard household electricity rate of $0.14 kWh was reached.

Figure 6.4 illustrates the results of shifting costs of electricity being sold to the grid on the net present value of the tidal turbine over varying design lives. For a 20-year design life, a discrepancy of $15.7 million was estimated for grid rates between $0.20/kWh and $0.14/kWh. As the design life of the project decreases, this discrepancy also decreases. With only a 5-year design life, this difference between $0.20/kWh and $0.14/kWh is only $7 million.

6.4 Discussion

This analysis has indicated that a tidal turbine would be able to satisfy the energy demands of a small community in Nova Scotia with a capital cost of $7.9 million and annual operations and maintenance costs of $6300. This in turn will make a profit of $37.7 million dollars over the 20-year lifespan of the turbine given a 3.7 MW capacity. Although the optimization analysis indicated that a much greater capacity could be achieved, this would be infeasible from a logistics perspective. The dimensions of a turbine with a 23.7 MW capacity would be astronomical and could lead to detrimental effects to sea life. Further, this type of structure would require intensive maintenance during which the plant would be completely shut down. Therefore, it is suggested that an array type design be created in the implementation of the project. The array type design would have the same apparent diameter as the optimized diameter we propose for the single turbine in this design. However, this would ensure that maintenance could be performed on a single turbine while the others continue to generate electricity and revenue for the plant and also minimize disruptions to the grid. This will further ensure that the return period is minimized as the amount of time the turbines are generating power will be maximized. The array type design would have all the turbines located at the optimum depth in a row to ensure that the maximal velocity can be extracted by each turbine. An array type design could also ensure that power could reach both sides of the passage and the transmission lines could connect to the mainland in multiple locations. Although this would further complicate the design, it would also be a way in which to optimize power generation during natural disasters or system failure. That is, failure of the entire system can be avoided if not all the energy is sent to a single transmitter that fails, but instead, the other transmitters would be able to continue to work while the other is being repaired. This proposed system would ensure the reliability and sustainability of the power system and provide greater peace of mind to the consumers in Nova Scotia. The scope of this work does not include the costs of transmission of energy to the communities. As a result, the overall revenue gained over the 20-year lifespan may be overestimated and the return period for the project may be underestimated. The revenue associated with the turbine is also highly dependent on the fluctuating costs of electricity in the area.

This analysis was based on the price consumers pay for energy but would need to include the greater variation in energy prices in the future in order to better predict the financial benefits of the project.

Furthermore, as the goal was to create a preliminary design of a turbine and its feasibility, the bearings and transmission design was not included. As a result, the shear depth of the turbine anchor could not be included in this analysis and the actual optimum depth may not have been established. Therefore, the turbine placement may need to be evaluated depending on the sediments lining the ocean floor. Additionally, a stability analysis may need to be conducted to determine the feasibility of constructing such a turbine within the Minas Passage. Lastly, the analysis did not consider in detail the environmental costs or impacts of the project. Current research to date has not extensively examined these issues either. Thus, it is important before crossflow turbine projects are implemented in any capacity to conduct a thorough environmental assessment in order to limit the damages to sea life and the future costs associated with remediating these areas. Overall, this analysis provided a preliminary sizing of the infrastructure that is required and that can be supported by the Minas Passage in terms of crossflow turbines. The optimized dimensions for the project prove that it could be very financially beneficial to the surrounding communities if such a project was adopted. Not only would it provide energy to the community, but also provide a means of income. Given further studies, tidal power could be a great solution to diversify our energy production away from fossil fuels.

6.4.1 Study Limitations

For the purposes of this study, a series of assumptions and simplifications were made during the design and optimization of the tidal turbine. These simplifications fall into three broad categories: simplifications in turbine design, simplifications made in optimization, and logistical simplifications. With respect to the design of the turbine, it was assumed that Betz theory applies. This theory is a simplified analysis of turbine power generation in that it does not consider the impacts of the surrounding bed or wall geometries and only considers the flow passing through a turbine. As an extension, this study recommends the usage of different theories for power extraction by a turbine. For example, Linear Momentum Actuator Disk Theory (LMADT) can be used as an alternative approach to estimating power generation from a fence of tidal turbines. The full LMADT theory is able to integrate in surface dynamics through incorporating in the Froude number and incorporation of the blockage ratio and is derived from Bernoulli's equation and conservation of momentum [14, 15]. The design of this turbine was further simplified by only considering 1-dimensional water flow. It should be noted that in reality, tidal velocities can vary in direction, and therefore, the power generation capacity estimated from this study may be an overestimate. Turbines are commonly employed in underwater operations for this reason as they can generate electricity regardless of the direction of water flow; however, they tend to have lower efficiencies than traditional turbines.

In terms of optimization, various simplifications and assumptions were made such that feasible solutions can be found. Many of these assumptions were based on the constraints used in the model. For instance, it was assumed that velocities of the water upstream of the turbine are constant for a given depth and that these velocities do not change with time. Additionally, it was assumed that the turbine wake speed (u_{t4}) was a third of the upstream tidal velocity. This assumption was based on the recommendation of Karsten et al. [14, 15]. A final set of simplifications were made with respect to the economic analysis conducted in this study. It was assumed that the applied discount rates (6–12%) are constant over the entire design life of the project, and no taxes or inflation was considered. Additionally, operations and maintenance costs were assumed constant over time and were adapted from the operations and maintenance costs of a small wind turbine farm. The logistics of the implementation of the turbine, and the structural considerations needed to secure the turbine the seabed were not part of the scope of this project and were therefore not considered. It is recommended that future studies consider the feasibility of implementation of the designed turbine.

Mechanical properties, such as density and viscosity, must be considered when choosing and configuring components that interact directly with the fluid. The solution needs to be pumped to the top of the absorber column. For a given volume, fluids that have higher densities have more mass and thus require more energy and power to be lifted. Similarly, a larger amount of force and energy is required to pump viscous fluids through a given gauge of pipe. Mechanical properties also affect reaction rates. Fluids that have higher viscosities tend to have droplets that are larger [17]. Larger droplets mean that the surface area to volume ratio is lower. This means that for a given volume of dripping fluid, the contact area it has with flue gas is lower. Since the CO_2 has to cross the surface before becoming absorbed by a droplet, a smaller surface area decreases the rate of absorption. Zhang et al. [7000] trained both GRNN and BPNN neural networks to predict the density and viscosity of potassium lysinate and monoethanolamine solutions, based on solute concentrations and the temperature. Each neural network had at least four inputs: the concentration of potassium lysinate, the concentration of monoethanolamine, the sum of the concentrations of both solvents, and the temperature. The inclusion of the total solvent concentration as an input is redundant since the exact same information is available to neurons by increasing the weights of individual concentrations by the same value. The inclusion of this unnecessary input adds more weights and parameters to the model, increasing the computational demand of training. Four datasets were used for the test and training data. Three datasets, which contained potassium lysinate only solutions, had diverse solvent concentrations, temperatures, densities, and viscosities. The one dataset that used solutions of sodium lysinate mixed with monoethanolamine had relatively limited variation in the concentrations of either solvent, the temperature, the density, and the viscosity. This creates discontinuities in the regions of the input space where the model can be used. To avoid this problem, the model should have been focused on predicting the physical properties of solutions with only sodium lysinate solutions dissolved within them. For the density predictions, the GRNN network, with Gaussian activation functions, had the best performance. This model was accurate, with a small mean square error of less than 1% of every possible value and 2.7% of the range. In contrast, for the viscosity predictions, a single hidden layer BPNN network with 2 hidden neurons had the best performance. The author did not mention the activation function that was used in each layer. Overall, the models for the viscosity were far from precise. At best, they had a root mean square error of 0.24 ⍰/mPa s for typical values of 2 ⍰/mPa s. This means that neither the GRNN nor the BPNN with the given activation function was a good model for viscosity.

6.4.2 Environmental Impact

Although this study did not take into consideration the environmental costs or benefits that would be incurred as part of this project, it is, nevertheless, important to note potential adverse environmental affects as a direct result of this project. Several impacts can arise as a result of the implementation of tidal turbines. These include static, dynamic, chemical, acoustic, electromagnetic, and cumulative effects [28]. There is also the issue of energy removal from the system. These effects can cause consequences at extensive distances upstream and downstream of the turbine [10]. Static effects are those from the physical structure of the turbine in water and the dynamic effects are a result of the movement of the turbine blades in the water, where faster blade speeds are expected to have a greater impact on marine life than slower blade speeds [37]. These dynamic effects are coupled with acoustic effects, which are a result of the vibrations of the turbine. Students and faculty at the University of Washington have identified high levels of noise form a tidal turbine when in operation [11]. Chemical effects arise from the construction of the project wherein lubricants or paints may leach into the ocean. These chemicals can be toxic to both marine life and humans consuming fish or using the ocean for recreational purposes [37]. The electromagnetic effects arise from the distribution of energy generated. Although most of the transmission lines run under the seabed, transmission lines must run from the turbine itself to the seabed and risk exposure to aquatic life. All of the above-mentioned stresses lead to long-term cumulative effects in which the effects become highly coupled and develop in the same geographical location, leading to greater effects to aquatic life.

The current main priorities are to better understand the static effects on benthic invertebrates, the structure of the water column, and the disruption to migrating species during device maintenance; the likelihood of marine life coming into contact and being harmed by the moving parts; how to reduce the leaching of paints and lubricants, the resuspension of pollutants in disturbed sediments, and the potential for fuel spills; the acoustic effects from pile driving, the tonal effects from moving parts, and behavioural changes in marine life due to noise creation from the turbines; the creation of electromagnetic fields and its impacts on behavioural adaptations of marine life; and the difficulties in predicting, detecting, and correlating the impacts of tidal turbines to these effects [28]. Crossflow turbines, as discussed here, have not been well studied in these areas though as there has not been widespread implementation globally. Thus, any tidal turbine being installed will be constructed under great environmental risk and will aid in answering the many gaps that currently exist in the literature regarding this topic.

6.5 Conclusion

This chapter presented the design of a crossflow turbine in the Minas Passage in the Bay of Fundy to meet the demand of a community of 5000 people in Nova Scotia. The minimum capacity required was 2.7 MW. The turbine design was carried out similar to the design of a wind turbine using the Betz theory. Optimization was conducted to maximize the power generation within the depth constraints of the channel by maximizing the economic value of the electricity generated and subtracting the costs of the turbine. The results show that a max power generation capacity of 23.2 MW can be achieved with a 56.8 m diameter turbine. To meet the demands of 5000 people, a turbine diameter of 45 m placed 75 m below the water surface was selected. This turbine design would be rated to 5.3 MW and therefore meet and exceed the required 2.7 MW. Economic analysis was also conducted, which revealed that based on an electricity selling cost of $0.20/kWh, the capital cost of $7.9 million, and the annual operations and maintenance cost of $6300, the payback period for the power plant will be 1.28 years and the profit over a 20-year lifetime would be $44.7 million. In addition, a sensitivity analysis was conducted to compare how the system would perform for different electricity prices over a 20-year design life, and the difference in profits between $0.14/kWh and $0.20/kWh after a 20-year life was found to be $15.7 million. The results of this study suggest the need to investigate the use of a turbine farm consisting of many smaller turbines to achieve a similar power output. This would have several advantages in terms of practicality, including ease of manufacturing and installation, less potential environmental disruption, avoidance of total plant shutdown for maintenance on one or a subset of turbines, thus minimizing disruption to the grid, and the possibility of having several points of transmission to cut down on transmission costs to multiple small communities surrounding the Minas Basin.

Simplifications in the turbine design and analysis were made, including the use of Betz theory which does not take into account the effect of the blockage ratio. As a result, assumptions were made about the velocity of flow through the turbine that would have resulted in maximum power generation, when in reality, this may not be the case. Taxes and inflation were also not considered in the economic analysis. It is recommended that detailed analysis be conducted in order to assess the feasibility. In addition, analysis of the environmental impact of the power plant has been carried out. Overall, the analysis presented demonstrates the potential for crossflow tidal-stream energy generation in the Minas Basin of the Bay of Fundy, as the energy production capacity well exceeded the minimum requirement to power a small community and economic returns exceeded expectations. It is expected that in conducting detailed analysis of a turbine farm, which takes into account the blockage ratio, the result would still result in a design which meets the requirements. With proximity to many small communities with modest power requirements, the Minas Basin is a great location to employ this power plant and supply renewable power to Canadians, thus aiding the transition away from a fossil fuel energy economy.

References

1. G. Allan, M. Gilmartin, P. McGregor, K. Swales, Levalized costs of wave and tidal energy in the UK: Cost competitiveness and the importance of "banded" renewables obligations certificates. Energy Policy **39**(1), 23–39 (2011). https://doi.org/10.1016/j.enpol.2010.08.029
2. Y.H. Bae, Lake Sihwa tidal power plant. Ocean Eng., 454–463 (2010)
3. A. Betz, Wind-Energie und ihre Ausnutzung durch Windmühlen. Göttingen Vandenhoeck & Ruprecht 1926 [Ausg. 1925] (1926)
4. Canada Energy Regulator, Provincial and territorial energy profile – Nova Scotia (2020). Retrieved from https://www.cer-rec.gc.ca/en/data-analysis/energy-markets/provincial-territorial-energy-profiles/provincial-territorial-energy-profiles-nova-scotia.html#:~:text=In%202017%2C%20annual%20electricity%20consumption,was%20residential%20at%204.5%20TW
5. C.A. Consul, *Hydrodynamic Analysis of a Tidal Cross-Flow Turbine* (Doctoral diseration). (Oxford University, 2011), p. 216
6. A. Cornett, N. Durand, and M. Serrer, 3-D modelling and assessment of tidal current resources in the Bay of Fundy, Canada, in *3rd International Conference on Ocean Energy*. Bilbao (2010). Retrieved from https://www.researchgate.net/publication/259623495_3-D_Modelling_and_Assessment_of_Tidal_Current_Resources_in_the_Bay_of_Fundy_Canada
7. A.S. Det Norske Veritas, Fatigue deisgn of offshore steel structures (2011). Retrieved from https://rules.dnvgl.com/docs/pdf/DNV/codes/docs/2011-10/RP-C203.pdf
8. D. Fallon, M. Hartnett, A. Olbert, S. Nash, The effects of array configuration on the hydro-environmental impacts of tidal turbines. Renew. Energy **64**, 10–25 (2014). https://doi.org/10.1016/j.renene.2013.10.035
9. H. Hickey, Assessing the environmental effects of tidal turbines (2010, December 13). Retrieved from University of Washington: https://www.washington.edu/news/2010/12/13/assessing-the-environmental-effects-of-tidal-turbines/
10. Karsten, R., Culina, J., Swan, A., O'Flaherty-Sproul, M., Corkum, A., Greenberg, D., & Tarbottom, M. (2011a). Assessment of the Potential of Tidal Power from Minas Passage and Minas Basin
11. R. Karsten, D. Greenberg, and M. Tarbotton, *Assessment of the Potential of Tidal Power from Minas Passage and Minas Basin* (Offshore Energy Research Association, 2011b). Retrieved from https://oera.ca/research/assessment-potential-tidal-power-minas-passage-and-minas-basin
12. V.K. Khare, *Tidal Engery Systems- Design Optimization,and Control* (Elsevier, 2019)
13. S. Klaus, Financial and economic assessment of tidal stream energy – A case study. Int. J. Financ. Stud. **8** (2020). https://doi.org/10.3390/ijfs8030048
14. D.M. Kurniawati, Experimental investigation on performance of crossflow wind turbine as effect of blades number. AIP Conf. Proc. (2018). https://doi.org/10.1063/1.5024104
15. N.D. Laws, B.P. Epps, Hydrokinetic energy conversion: Technology, research and outlook. Renew. Sust. Energ. Rev. **57**, 1245–1259 (2016). https://doi.org/10.1016/j.rser.2015.12.189

16. Y. Li, and H.K. Florig, Modeling the operation and maintenance costs of a large scale tidal current turbine farm. *Oceans 2006* (Carnegie Mellon Electricity Industry Center, 2006a). doi: https://doi.org/10.1109/OCEANS.2006.306794
17. Y. Li, and K.H. Florig, Modeling the operation and maintenance costs of a large scale tidal current turbine farm. *OCEANS 2006* (2006b)
18. S. MacDougall, *Financial Evaluation and Cost of Energy* (Acadia Tidal Energy Institute, n.d.). Retrieved December 6, 2020, from https://tidalenergy.acadiau.ca/tl_files/sites/atei/Content/Reports/Module_7_Financial_Evaluation_and_Cost_of_Energy.pdf
19. A. Maia, E. Ferreira, M. Oliveira, L. Menezes, A. Andrade-Campos, Numerical optimization strategies for springback compensation in sheet metal forming, in *Computational Methods and Production Engineering*, (Woodhead Publishing, 2017), pp. 51–82
20. Natural Resources Canada, Electricity facts (2020a). Retrieved from nrcan.gc.ca/science-data/data-analysis/energy-data-analysis/energy-facts/electricity-facts/20068
21. Natural Resources Canada, Renewable energy facts (2020b). Retrieved from What is renewable enegy?: https://www.nrcan.gc.ca/science-data/data-analysis/energy-data-analysis/energy-facts/renewable-energy-facts/20069#L7
22. Nova Scotia Power, Tariffs (2019, April 29). Retrieved from https://www.nspower.ca/docs/default-source/default-document-library/nspowertariff.pdf?sfvrsn=fed97849_0#:~:text=set%20out%20below.-,ENERGY%20CHARGE,for%20all%20additional%20kilowatt%20hours
23. B.V. Polagye, *Environmental Effects of Tidal Energy Development* (U.S Department of Commerce, Natioanl Oceanic and Atmospheric Administration, 2010)
24. B. Polagye, B. Van Cleve, A. Copping, and K. Kirkendall, *Environmental Effects of Tidal Energy Development* (2010)
25. P. Radomirovic, Deisgn of the underwater structures, in *Contemporary achievements in civil engineering* (Subotica, 2017). doi:https://doi.org/10.14415/konferencijaGFS2017.011
26. A.M. Redden, M.J. Stokesbury, J.E. Broome, F.M. Keyser, A.J. Gibson, E.A. Halfyard, . . . R. Karsten, *Acoustic Tracking of Fish Movements in the Minas Passage and Force Demonstration Area: Pre-turbine Baseline Studies (2011–2013)* (Offshore Energy researcg Association and the Fundy Ocean Research Centre for Energy, 2014). Retrieved from https://tethys.pnnl.gov/sites/default/files/publications/ACER_Publication[2].pdf
27. R. Sharma, B. Glemmestad, On generalized reduced gradient method with multi-start and self-optimizing control structure for gas lift allocation optimization. J. Process Control **23**(8), 1129–1140 (2013)
28. Simec Atlantis Energy, Turbines & engineering services (n.d.). Retrieved from Tidal Turbines : https://simecatlantis.com/services/turbines/
29. Statistics Canada, Population and Dwelling Count Highlight Tables, 2016 Census (2016a). Retrieved from https://www12.statcan.gc.ca/census-recensement/2016/dp-pd/hlt-fst/pd-pl/Comprehensive.cfm
30. E.A. Sudderth, K.C. Lewis, J. Cumper, A.C. Mangar, and D.F. Flynn, *Potential Environmental Effects of the :eading Edge Hydrokinetic Energy Technology* (Department of Energy, Washington, 2017). Retrieved from https://rosap.ntl.bts.gov/view/dot/12471/dot_12471_DS1.pdf?
31. U.S. Department of Commerce- National Oceanic and Atmospheric Administration, Where is the highest tide? (2018)
32. C.R. Vogel, R.H. Willden, Improving tidal turbine performance through multi-rotor fence configurations. Marine Sci. Appl. **18**, 17–25 (2019). https://doi.org/10.1007/s11804-019-00072-y
33. F.-W. Zhu, L. Ding, B. Huang, M. Bao, J.-T. Liu, Blade design and optimization of a horizontal axis tidal turbine. Ocean Eng. **195** (2020). https://doi.org/10.1016/j.oceaneng.2019.106652
34. Zhang, Z., Li, H., Chang, H., Pan, Z., & Luo, X. (2018). Machine learning predictive framework for CO2 thermodynamic properties in solution. Journal of CO2 Utilization **26**, 152–159

Chapter 7
Renewable Energy Integration for Energy-Intensive Industry to Reduce the Emission

Cheng Seong Khor, Ali Ahmadian, Ali Almansoori, and Ali Elkamel

7.1 Introduction

7.1.1 Background and Motivation

Depleting fossil fuel resources and growing environmental concerns particularly due to climate change has thrusted renewable energy as an option for sustaining energy supply in meeting demand for energy-intensive industries such as oil and gas. One such solution is to adopt renewables such as solar energy for hydrogen production in crude oil upgrading process. Nearly 95% of worldwide energy is currently produced from fossil fuel resources, e.g., natural gas, hydrocarbon, and coal, which have been depleted and have a tremendous environmental impact [1].

According to recent projection published in World Oil Outlook 2045, global energy demand is expected to increase by 60% from 256 to 410 million barrels of oil

C. S. Khor
Department of Chemical Engineering, Universiti Teknologi PETRONAS, Tronoh, Perak, Malaysia

Centre for Systems Engineering, Universiti Teknologi PETRONAS, Perak Darul Ridzuan, Malaysia

A. Ahmadian (✉)
Department of Electrical Engineering, University of Bonab, Bonab, Iran

Department of Chemical Engineering, University of Waterloo, Waterloo, ON, Canada

A. Almansoori
Department of Chemical Engineering, Khalifa University, Abu Dhabi, United Arab Emirates
e-mail: ali.almansoori@ku.ac.ae

A. Elkamel
Department of Chemical Engineering, University of Waterloo, Waterloo, ON, Canada

Department of Chemical Engineering, Khalifa University, Abu Dhabi, United Arab Emirates
e-mail: aelkamel@uwaterloo.ca

A. Ahmadian et al. (eds.), *Carbon Capture, Utilization, and Storage Technologies*, Green Energy and Technology, https://doi.org/10.1007/978-3-031-46590-1_7

equivalent per day in 2040 in which fossil fuels will remain the primary source for supplying this rising demand [2]. However, over recent decades, there has been an increase in concern over the impact of fossil fuels on the environment [3]. Continuous production of crude oil will eventually result in the depletion of conventional oil reserves. To meet the demand, the oil and gas industry has resorted to extract and refine crude oils from unconventional heavy reserves such as oil sands [4], and the development of such processing facilities is expected to rise. These unconventional reserves require more energy to both produce (extract) and process with consequent increased energy consumption. Additionally, the higher energy consumption possibly leads to greater environmental impact such as undesired greenhouse gas (e.g., CO_2 emissions). The situation calls for the oil and gas industry to consider more renewable energy use.

7.1.2 Role of Solar Energy for Industrial Processes

Solar energy has been used for the past 40 years or so to supply electricity to off-grid communities via photovoltaic systems (PV) which offers a viable alternative if electricity is not accessible [5–7]. Another important application of solar energy is to supply thermal energy for the process and its affiliated industries mainly for steam and/or heat generation [8, 9]. A pioneering example includes a solar thermal facility constructed by the oil company Chevron (through its subsidiary of Chevron Technology Ventures) and BrightSource Energy, which produces steam through solar concentrators and in so doing assists Chevron to reduce fossil fuel use in the overall production or manufacturing activity [3, 10].

This chapter focuses on potential solar energy utilization to produce hydrogen sustainably for the oil and gas industry [11]. The application of solar energy concentrators is prospective particularly for heavy oil upgrading process such as that encountered in oil (or tar) sands mining [12]. Substantial heavy oil resources are often found in locations with high solar irradiance for which solar energy use can be efficient and cost effective. Additionally, solar energy can be directly used to provide high temperature heat to meet the requirements of such highly endothermic reactions. However, existing technologies are largely under development with no industrial scale applications attempted to date [3, 13].

7.1.3 Heavy Crude Oil Upgrading with Hydrogen

Environmental regulations require removing impurities (mainly sulfur) from end-use petroleum products. Heavy crude oils require special techniques for extraction and recovery; they cannot be processed directly in current petroleum refineries. Therefore, oil companies usually construct upgraders near such oil production fields to convert the extracted oil, commonly called bitumen, into synthetic oil using hydrogen. Based on current practice, refineries mostly supply the required hydrogen through conventional steam methane reforming [4, 14]. In the upgrading process, heavy oils extracted from production fields is first fed to a diluent recovery unit (DRU) to separate and recover the diluent use in its processing. DRU operates at atmospheric conditions and produces two other outlets besides diluent, namely, light gas oil (LGO) and a bottoms stream that contains heavy hydrocarbon components. The bottoms stream is then cracked into smaller components in a thermal cracking process unit (called primary cracker) where lighter components of naphtha, light gas oil, and heavy gas oil are obtained. An optional secondary cracker can also be employed to crack the residuals from the primary cracker further to give more yields. Thereafter, the outlets or products are sent for hydrotreating, which is the main hydrogen user for upgrading to remove sulfur, nitrogen, aromatics, and other impurities. Hydrogen demand for the upgrading chiefly depends on the amount of impurities to be removed [23].

7.1.4 State-of-the-Art Hydrogen Production Methods

There is much active research that focuses on renewable hydrogen production [15]. Three main methods to produce hydrogen via solar energy include photochemical, thermochemical, and electrochemical pathways as briefly summarized in Table 7.1. A photochemical-based process uses sunlight for water hydrolysis to produce hydrogen. Although it can be carried out with only sunlight heat, this technique is considered not practical because over 2000 °C is required in the reaction to dissociate water. Despite numerous efforts to improve its practicality, the photochemical pathway remains under research and not commercialized [16].

Electrochemical and thermochemical pathways offer better prospect to produce hydrogen from solar energy. An electrochemical process, which involves water electrolysis, is a mature technology besides allowing hydrogen to be produced with low greenhouse gas emissions. But it is not cost–effective compared to the other two technologies [1, 17, 18]. Two methods are available to produce hydrogen electrochemically: (i) by harvesting solar energy via photovoltaic technology to generate electricity directly from sunlight [20]; (ii) by concentrated solar energy stored in thermal energy storage for use to generate electricity in a steam turbine cycle. The generated electricity is supplied to electrolyzers to produce hydrogen. The

Table 7.1 Comparison of hydrogen production methods [15]

Method	Process (reaction)	Technoeconomic assessment
Photochemical	Water hydrolysis	Temperature ≥ 2000 °C (not practical); not commercialized [16]
Electrochemical	Water electrolysis	Low emissions but not cost-effective; medium maturity (~4% market share) [1, 17–19]
Thermochemical	Steam methane reforming	Commercialized, high maturity (~96% market share) [19]

latter method also offers additional benefit in that an excess of stored thermal energy can be supplied to a crude oil upgrader as lower-grade heat [3].

The thermochemical pathway can use solar energy to supply the heat needed for endothermic catalytic hydrocarbon transformation reactions such as cracking and steam reforming for hydrogen production. A thermochemical process with fossil fuel feedstock is potentially the most well-developed and commercially exploited technology to produce hydrogen. It is reported that an estimated 96% of global hydrogen production is derived from fossil resources while the remaining is from water electrolysis. Currently, steam methane reforming is the most widely used thermochemical process, accounting for about 48% of total hydrogen production in which natural gas is reformed with steam to produce syngas (which is made up mainly of hydrogen) [1, 19].

Although several promising solar-aided technologies are available to produce hydrogen (particularly solar thermal reforming of methane for hydrogen and syngas generation), they remain expensive and a few still need further technical developments [21–23]. In the foreseeable future of transitioning to renewable energy-based technologies, hydrogen production with competitive cost can be achieved only by steam reforming of methane integrated with solar energy. A solar energy concentration system is commercially available to produce thermal energy up to 1000 °C as heat source for the endothermic reforming reactions. A solar receiver reactor can be used that is directly exposed to concentrated sunlight to receive and transfer thermal energy as reactants heat. Projects have been carried out to develop solar receiver reactors that are upscaled and tested in actual operations. In general, there is significant progress achieved with solar-aided steam methane reforming including at pilot plant demonstration scale. Hence, combining solar energy with conventional steam methane reforming is plausible to produce the hydrogen required to upgrade crude oil extracted from tar sands sustainably. However, these technologies still need public and financial support for market deployment [21].

Several efforts exist that exploit solar energy to produce hydrogen supply for the oil and gas industry, but they have not been attempted at industrial scale. Consequently, this chapter investigates several promising pathways of such solar energy–based hydrogen production, namely, solar steam methane reforming using volumetric receiver reactor (SSMR-VRR), solar steam methane reforming using molten salt as a heat carrier (SSMR-MS), and solar thermal power generation coupled with water electrolyzer (STP-WE). We propose conceptual design for these pathways and evaluate them from technical, economic, and environmental aspects to assess their suitability to be employed at industrial scale.

7.2 Modeling Framework

Several process alternatives involving solar energy–based hydrogen production are studied in this chapter. The alternatives considered include solar steam methane reforming of natural gas using volumetric receiver reactor (SSMR–VRR), solar steam methane reforming of natural gas using molten salt as a heat carrier

(SSMR–MS), and solar thermal power generation using water electrolysis (STP–WE) for crude oil upgraders as shown in Fig. 7.1. These alternatives are compared with the conventional steam methane reforming process (CSMR) from technical, economic, and environmental standpoints.

To perform the comparison, we employ the System Advisor Model (SAM) as a platform to model and design the solar components of the process (e.g., solar power concentrators) [24]. Process simulation for crude oil upgrading is carried out using Aspen Plus to determine hydrogen demand requirements [25]. Data for this process considers feedstock of 150,000 barrel per day (bpd) of bitumen [26].

Figure 7.2 presents a simplified flow scheme used to develop the crude oil upgrading process simulation model. A crude oil upgrader typically consists of a distillation unit and a residual oil processor, e.g., vacuum distillation unit and delayed coker. The main function of upgraders is to convert bitumen into synthetic crude oil for subsequent processing in conventional refineries. The synthetic crude oil contains light components such as naphtha, diesel, and heavy gas oil. The upgrader model also includes a hydrogen treater to recover hydrogen from the hydrotreater purge gas. A hydrogen demand of 2577 kmol per hour is estimated for the bitumen upgrader [26].

Based on the simulation results, four schemes proposed on alternative hydrogen production methods are considered as follows [1, 21]:

Scheme 1 (base). Conventional Steam Methane Reforming of natural gas process (CSMR).

Scheme 2. Solar Steam Methane Reforming of Natural Gas Using Volumetric Receiver Reactor (SSMR-VRR): A solar volumetric receiver reactor is used to

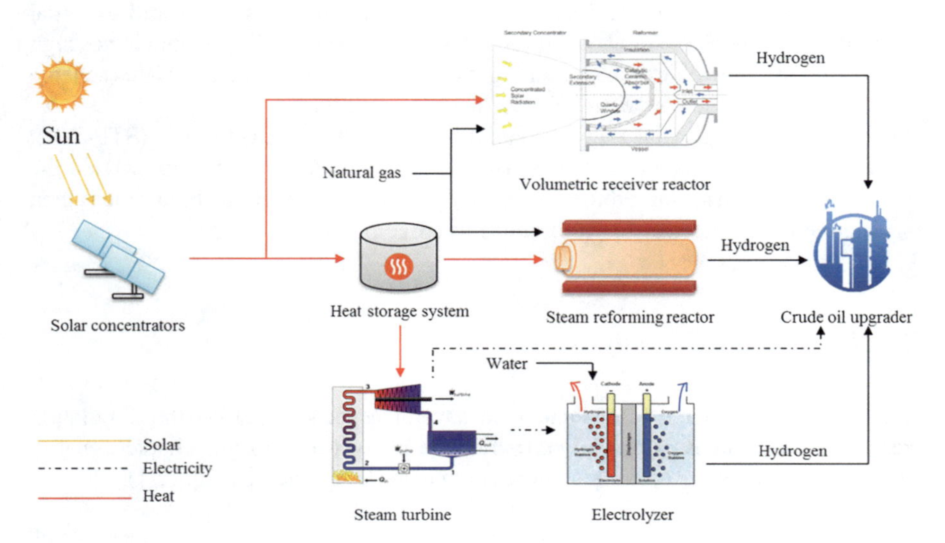

Fig. 7.1 Hydrogen production alternative schemes using solar energy considered in this chapter [27, 28]

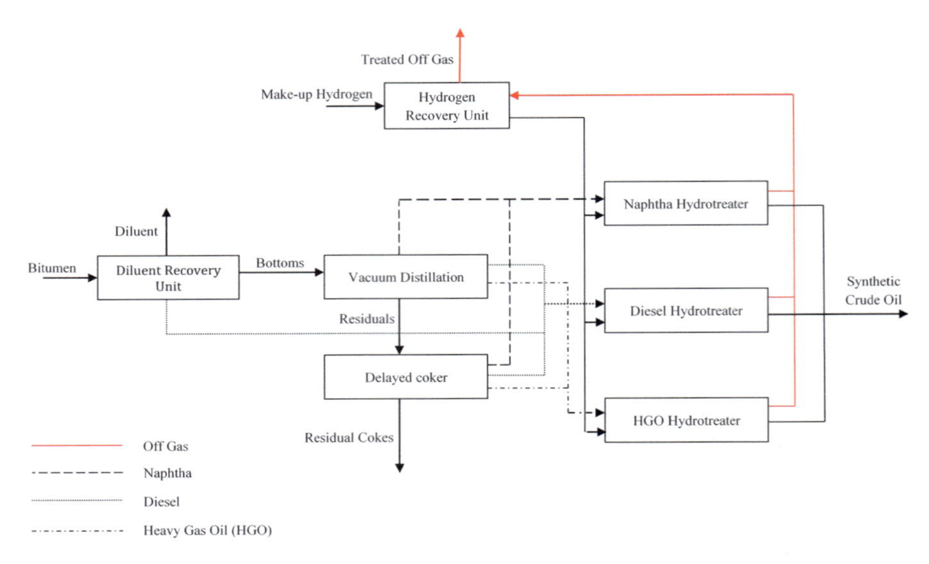

Fig. 7.2 Flow scheme (simplified) for crude oil upgrading process [27, 28]

directly absorb concentrated sunlight and produce syngas. However, a conventional reformer (CSMR) is needed for the rest of the operation due to limitation that allows it to operate only 2000 hours per year.

Scheme 3. Solar Steam Methane Reforming of Natural Gas Using Molten Salt as a Heat Carrier (SSMR-MS): Molten salt is heated by using concentrated solar energy and its thermal energy is transferred to a steam reforming reactor to drive the reactions forward. However, a conventional reformer is still essential in this process as the molten salt provides heat up to only 500 °C that is not high enough to achieve desirable natural gas conversion. An advantage of this scheme is that duty of the conventional reformer can be reduced.

Scheme 4. Solar Thermal Power Generation Using Water Electrolysis (STP-WE): This solar power to gas pathway uses concentrated solar energy to provide the heat required in steam turbine. Electricity generated from the steam turbine is then supplied to an electrolyzer to produce hydrogen.

7.2.1 Technical Analysis

Each proposed scheme is evaluated on several aspects based on the simulation results. Process analysis can be carried out in many ways. In this chapter, energy efficiency is mainly used for technical comparison as defined by Eq. (7.1):

$$\text{Energy efficiency} = (\text{Total energy input})/(\text{Total energy output}) \times 100\% \tag{7.1}$$

where total energy input is a summation of different energy carriers supplied to the process (e.g., heat, electricity, solar, and fuel type based on lower heating value [LHV]). Similarly, total energy output is a summation of all energy carriers leaving the process, which is calculated based on LHV of produced hydrogen.

7.2.2 Economic Analysis

Since each proposed hydrogen production pathway utilizes different methods and technologies, a means of equally comparing the cost for all the schemes is adopted as follows. For economic evaluations, a levelized cost of hydrogen production (LCHP) is used to compare the cost of producing an equivalent amount of hydrogen across all schemes. We calculate LCHP ($/kg) based on an amortized total cost and the total amount of hydrogen produced during the lifetime of the system as given by eq. (2).

$$\text{LCHP} = \frac{1}{F_H \times h \times n} \sum_{y=1}^{n} \frac{C_{\text{Capital}} + C_{\text{Production}}}{(1+r)^{y-1}} \tag{7.2}$$

where C_{Capital} = total annualized capital cost, $C_{\text{Production}}$ = total production cost of hydrogen (including operational, maintenance, and fuel costs), F_H = hydrogen production rate (kg/hour), which is equal to hydrogen demand for crude oil upgrading, n = project lifetime, and r = discount rate. The annual system operation time is assumed as $h = 8000$ h.

The four schemes are analyzed over an expected 20-year lifecycle of a hydrogen production facility. This analysis allows us to compare and identify processes that are significantly different from each other based on LCHP.

In this work, we use the size and capacity of each equipment to estimate its purchasing cost. We develop a cost estimation method based on Lang's factorial method [29]. The purchasing cost of solar-related equipment is estimated using the method available in the SAM software [24] while those of the electrolyzers and its components are taken from the H2A model developed by the US National Renewable Energy Laboratory (NREL) [30].

7.2.3 Environmental Impact Analysis

To address environmental impact reduction, benefits of using solar energy for each of the hydrogen production schemes are analyzed. We compute amount of CO_2 emissions and removal in flue gas generation and amine absorption processes.

7.3 Process Design Simulation Modeling

We carry out process simulation for all four schemes in Aspen Plus while that of the solar concentrators system design is performed in SAM. Details of simulation model for each scheme is discussed next.

7.3.1 Scheme 1 (Base): Conventional Steam Reforming of Natural Gas Process (CSMR)

Treated natural gas and steam are supplied as feed for steam methane reforming. The reformer high temperature (900 °C) is attained by combusting natural gas. The generated syngas leaving the reformer is cooled and enters a water-gas shift reactor to convert carbon monoxide (CO) into more hydrogen product. Syngas passes through a train of purification units to produce hydrogen to the required specifications starting with water separation (removal) using flash drum followed by amine gas treating to remove carbon dioxide (CO_2) from the product stream. Finally, the hydrogen produced is purified using pressure swing adsorption (PSA) to achieve purity up to more than 99.9 volume percent (vol%). However, only 90 vol% of the produced hydrogen is recovered from this unit. Off-gas from PSA is recycled as fuel to supply the reformer heat. Figure 7.3 shows the CSMR process flow.

7.3.2 Scheme 2: Solar Steam Reforming of Natural Gas Using Volumetric Receiver Reactor (SSMR-VRR)

The SSMR-VRR scheme is similar to the foregoing CSMR process with the addition of volumetric receiver reactor and solar concentrators systems. A key difference is that the PSA off-gas is recycled as reformer feed instead of fuel. This allows providing heat directly by concentrated solar energy that can reduce natural gas consumption (see Fig. 7.4).

7.3.3 Scheme 3: Solar Steam Reforming of Natural Gas with Molten Salt as Heat Carrier (SSMR-MS)

The third scheme involves solar steam reforming of natural gas using molten salt as a heat carrier (SSMR-MS). It is identical to the conventional process (CSMR in scheme 1). The difference lies in the addition of a pre-reformer unit that utilizes molten salt heated by concentrated solar energy as a heat carrier. The pre-reformer partially converts natural gas into hydrogen, thereby reducing heat duty required in

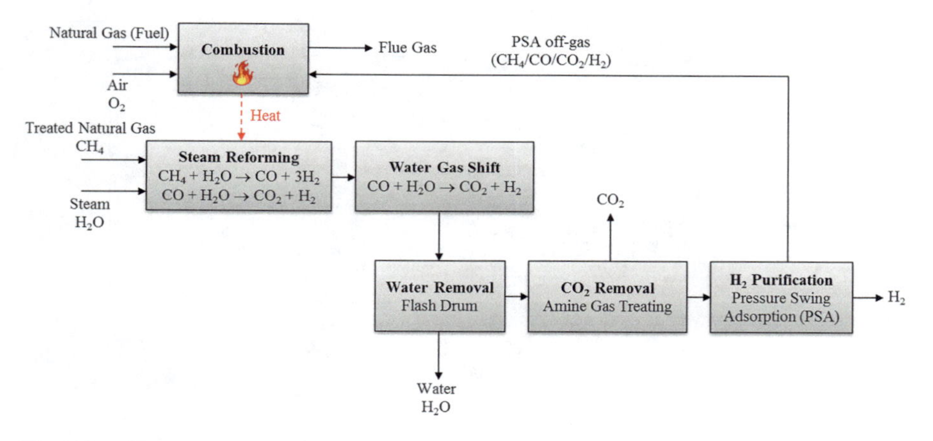

Fig. 7.3 Process schematic of conventional steam reforming of natural gas (CSMR)

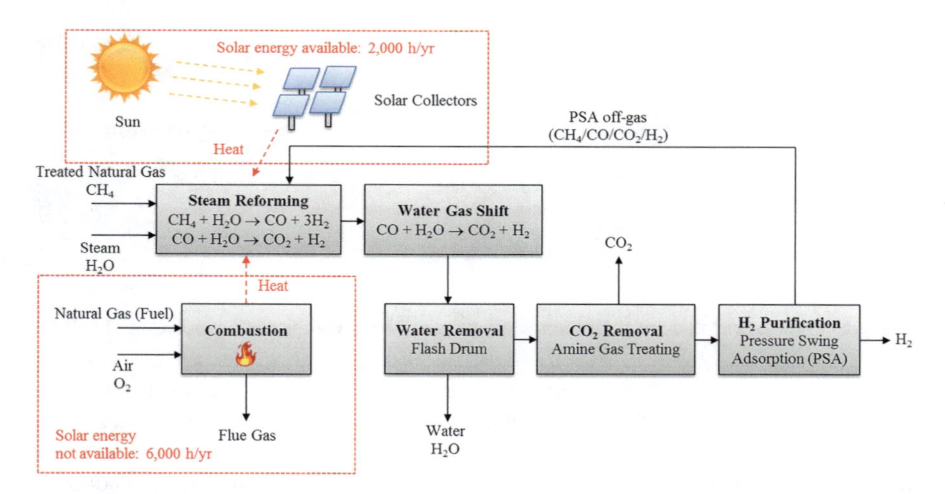

Fig. 7.4 Process schematic of solar steam reforming of natural gas using volumetric receiver reactor (SSMR–VRR)

the main reformer. The rest of the process remains the same as CSMR as shown in Fig. 7.5.

7.3.4 *Scheme 4: Solar Thermal Power Generation with Water Electrolysis (STP-WE)*

The fourth scheme of solar thermal power generation with water electrolysis (STP-WE) is different from the first three alternatives in terms of the hydrogen

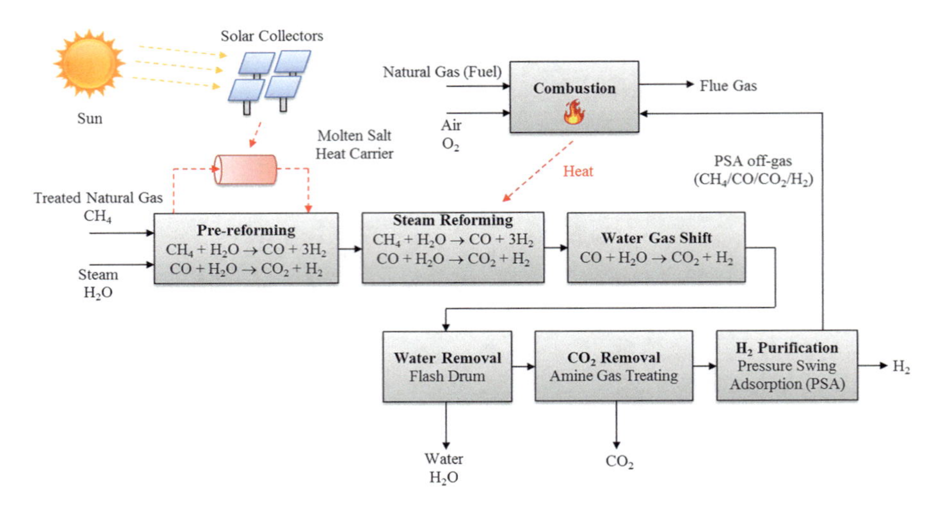

Fig. 7.5 Process schematic of solar steam reforming of natural gas using molten salt as a heat carrier (SSMR–MS)

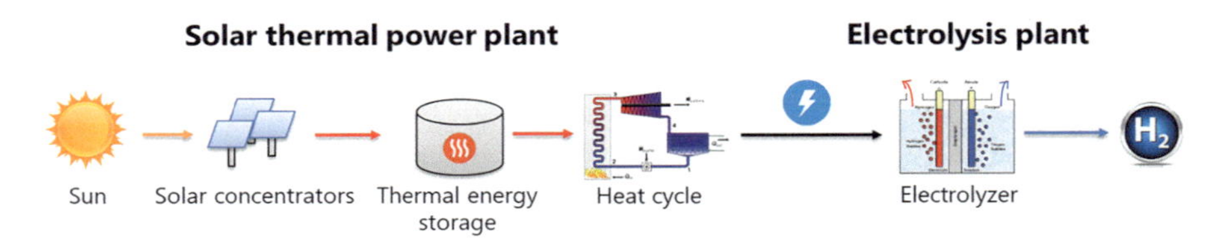

Fig. 7.6 Process schematic of solar thermal power generation with water electrolysis (STP–WE)

production method. STP-WE generates electricity through a heat cycle that uses steam turbine to receive heat from solar concentrators. The electricity generated is then used to produce hydrogen in the electrolyzer through electrolysis of water. The alkaline electrolysis technology is selected due to its higher hydrogen production capacity than the other three alternatives [31], for which we adopt an existing commercial electrolyzer (Norsk Hydro Atmospheric Type No. 5040) in this work [32]. However, using solar cells to generate electricity is relatively expensive as compared to other options; hence, this alternative is not considered further in this work (Fig. 7.6).

Flowsheeting for the first three schemes is carried out using the Aspen Plus process simulator. Based on a design target of hydrogen production rate of 2577 kmol/h, we conduct the simulation to determine the associated equipment size and other process requirements. The solar concentrator systems are designed via SAM by selecting solar tower technology that involves an operating temperature of up to 1000 °C, which is adequate to provide the heat required for steam reforming reaction [33].

7.4 Simulation Modeling Results

Aspen Plus simulation base data for the hydrogen production schemes are given in Table 7.2. The heat duty required for volumetric receiver reactor (in SSMR-VRR scheme) and the heat transfer in the molten salt heat carrier (in SSMR-MS scheme) are key design variables for the solar concentrator. For the STP-WE scheme, electrolyzer power consumption is a design target for the solar thermal power plant.

Table 7.3 summarizes comparison across all the hydrogen production methods considered in terms of fuel consumption, CO_2 emissions, solar system specifications, and land area requirement. Heliostats' reflective area size of STP-WE is very large as compared to those of SSMR-VRR and SSMR-MS. This incurs a large capital cost and land area to implement STP-WE which is challenging especially in land-scarce regions. Moreover, water consumption for STP-WE is the highest due to water electrolysis reaction requirement.

Table 7.2 Data for design simulation

Parameter	Value
Hydrogen production rate (F_H, demand)	2577 kmol/h (5154 kg/h)
Hydrogen purity	99.9 vol%
Maximum solar concentrator operating temperature	1000 °C

Table 7.3 Process simulation result summary

Parameter/variable	CSMR	SSMR-VRR	SSMR-MS	STP-WE
Natural gas consumption (kt/year)	150.35	129.85	132.36	–
Water consumption (kt/year)	455.95	417.19	454.16	462.14
Electricity consumption (GWh)	106.61	92.72	83.73	2218.01 (280 MW)[a]
CO_2 emissions (kt/year) (flue gas and removed CO_2)	411.25	355.19	356.83	–
Required heat duty for solar concentrators (MW)	–	65.95	43.79	–
Solar thermal energy design point (MW)	–	66.00	44.00	909.00[b]
Solar concentrators area (m^2)[c]	–	173,000.00	187,000.00	1,800,000.00
Land area required (acre)		281.00	297.00	3946.00
Solar tower height (m)	–	112.00	93.00	214.00
Thermal storage heat duty (MW)[d]	–	–	792.00	6816.00

[a]Design capacity of solar thermal power plant (for STP-WE scheme) is 140 MW that necessitates two of such systems

[b]Solar thermal energy design point of STP-WE design assumes thermal to electrical power conversion efficiency of 37% based on SAM

[c]Heliostats' (solar concentrators) area is designed based on direct normal irradiance of 950 W/m^2 (typical value for a location in Canada is used for solar irradiance

[d]Value determined assumes 15 h of heat storage

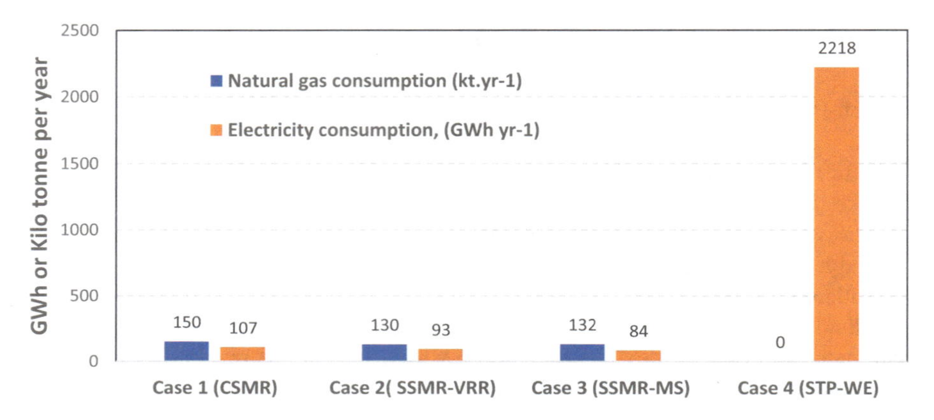

Fig. 7.7 Comparison between natural gas and electricity consumption across hydrogen production schemes

On the other hand, a noteworthy advantage of STP-WE is that it does not emit CO_2 during hydrogen production. The scheme also generates high-purity oxygen as a byproduct that can be used elsewhere in the refinery or sold for extra credits. However, STP-WE does not produce steam for crude oil upgrading. Another disadvantage is that it requires large solar-related equipment as indicated by Table 7.3.

A first important process parameter is natural gas consumption in each process as shown in Fig. 7.7. Solar energy use can reduce natural gas consumption rate in both the feed and the fuel. For SSMR-VRR scheme, a reduction of 14% compared to CSMR scheme is achieved and occurs mostly in the feed because the PSA off-gas is recycled to inlet whereas that of SSMR-MS scheme is of 12% in the fuel because of a lower heat duty in the main reformer. On the overall, both SSMR-VRR and SSMR-MS schemes can save an almost equivalent amount of natural gas consumption. Note that STP-WE scheme does not entail any natural gas use.

As also seen from Fig. 7.7, employing solar energy in steam reforming of natural gas reduces electricity consumption. Electricity used for compressing air decreases substantially for both SSMR-VRR and SSMR-MS. However, electricity consumed by the PSA off-gas compressor in SSMR-VRR increases by 10.4 GWh per year as compared to that of CSMR, because off-gas in the former needs to be re-pressurized to a higher level. This contrasts with SSMR-MS in which off-gas is recompressed to be used as fuel; thus, no extra electricity is needed.

On the overall, SSMR-MS entails a lowest electricity consumption rate. SSMR-VRR and SSMR-MS achieve a reduction of 13% and 21% in electricity consumption, respectively, compared to CSMR. Electricity consumption of STP-WE scheme is the highest due to electrolytic hydrogen production (which is generated through solar-aided steam turbines).

7.4.1 Technical Analysis

Energy efficiency of all the hydrogen production schemes considered (as computed by Eq. 7.1) is used as a basis for technical comparison as reported in Fig. 7.8. Overall energy input and output of all the schemes are displayed in Fig. 7.9. Energy content of natural gas and hydrogen are based on a lower heating value (LHV) of 5.003×10^4 kJ/kg and 11.996×10^4 kJ/kg, respectively.

STP-WE has the lowest energy efficiency as it involves more energy conversion steps besides that converting thermal energy to power entails low efficiency. Both SSMR-VRR and SSMR-MS schemes offer higher efficiency than CSMR due to energy loss through flue gas from combustion in CSMR. For SSMR-VRR and

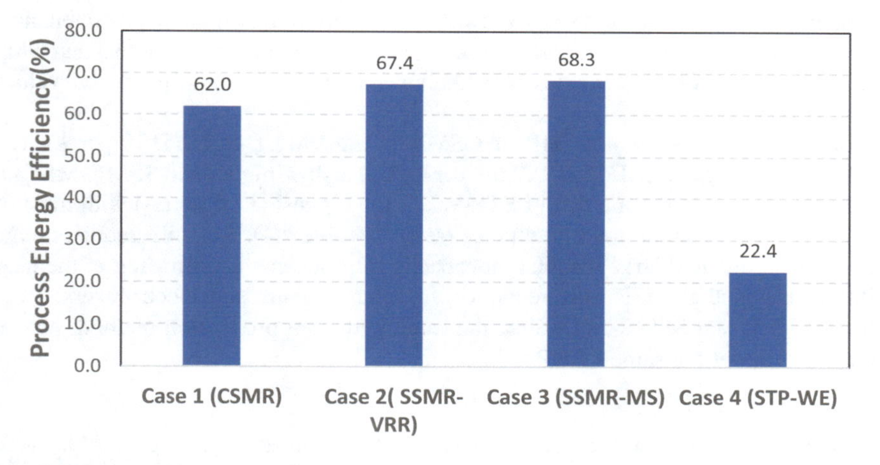

Fig. 7.8 Energy efficiency of hydrogen production schemes

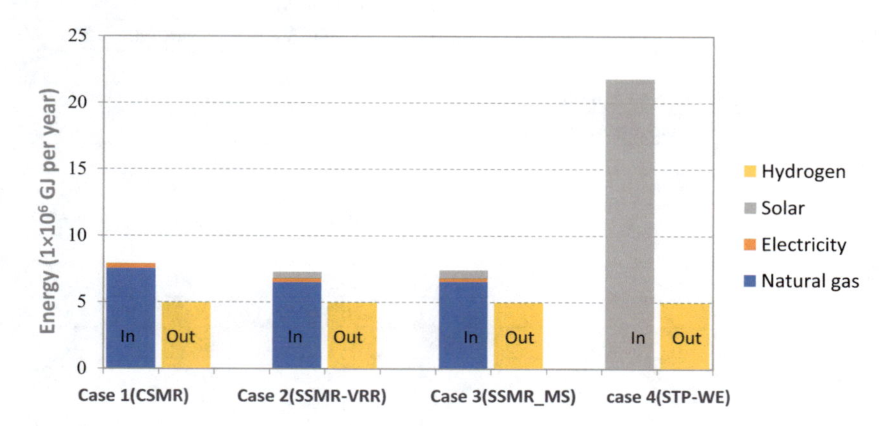

Fig. 7.9 Overall energy input and output of all hydrogen production schemes

SSMR-MS, solar energy is used to provide heat, thereby reducing heat loss through flue gas.

Figure 7.8 shows that SSMR-MS has highest energy efficiency at 10% greater than CSMR while that of SSMR-VRR is at 9%. In the former, heat is obtained from concentrated solar energy and exchanged through molten salt whereas in the latter heat is directly absorbed by the volumetric receiver reactor and used for steam reforming.

7.4.2 Economic Analysis

The levelized cost of hydrogen production (LCHP) requires considering capital investment cost, production cost, and hydrogen production rate. The production cost components comprise those of raw materials, utilities, operations, maintenance, and overhead. Table 7.4 reports the estimated cost components associated with LCHP for each of the schemes with their per unit of hydrogen produced values compared in Fig. 7.10.

The results show that LCHP of CSMR is the lowest at 1.7 US$ per kg H_2, followed by that of SSMR-VRR (2.5 $ per kg H_2) at 46% higher and SSMR-MS (3.8 $/kg) at 50% higher while that of STP-WE is the highest ($7.6/kg H_2) with most of the latter due to capital investment cost (refer to Table 7.4). STP-WE requires huge investment due to a large installation capacity; this scheme is promising in the near future provided there is cost reduction in solar systems and electrolyzers. Both SSMR-VRR and SSMR-MS utilize the same hydrogen production method; hence, they give about the same LCHP values.

Table 7.4 Summary of levelized cost of hydrogen production (LCHP) cost components associated with different hydrogen production schemes

	Value (in million $)			
Cost component	CSMR	SSMR-VRR	SSMR-MS	STP-WE
Capital investment				
Purchase cost of all process equipment	25.9	25.6	26.6	243.9
Purchase cost of all solar-related equipment	–	41	98.8	1853.2
Land	1.6	6.6	10.9	79
Working capital	22.2	60.2	113.3	202
Total	154.5	406.5	756.1	3962.8
Production cost				
Raw materials	24.7	20.5	24.8	2.7
Utilities	17.1	16.4	13	4.8
Operations	2.8	4.5	4.5	12
Maintenance	10.4	27.2	51.1	105.7
Overhead	5.6	14	25.4	44.8
Total	60.6	82.4	118.8	170.

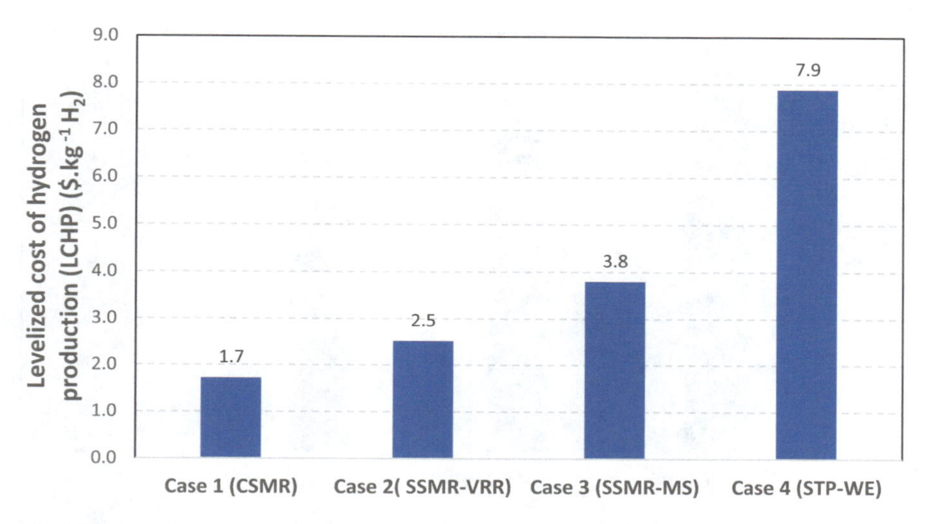

Fig. 7.10 Levelized cost of hydrogen production (LCHP) for hydrogen production schemes

7.4.3 Environmental Impact Analysis

Environmental impact of the schemes is considered in terms of total CO_2 emissions rate during hydrogen production (which includes CO_2 removed in amine absorption and flue gas processes) as shown in Fig. 7.11. CO_2 emissions correspond to natural gas consumption rate; i.e., lower fuel consumption leads to less CO_2 emissions. CO_2 emissions decrease notably for both SSMR-VRR and SSMR-MS in their combustion flue gas by 45.5 and 53.6 kiloton per year, respectively, which indicate the benefits of solar energy use for steam reforming of natural gas. Note that STP-WE does not entail any CO_2 emission as it beneficially uses renewable energy and water electrolyzer.

7.5 Modeling Results Summary

Figure 7.12 shows relative performance of all four schemes in terms of economic, environmental, and technical dimensions simultaneously. The thermochemical processes of SSMR-VRR and SSMR-MS hold more potential economically to produce hydrogen required for crude oil upgrading. Besides that, SSMR-VRR has lower LCHP than SSMR-MS (as discussed previously). SSMR-MS is the most energy-efficient scheme while STP-WE registers the lowest greenhouse gas emission rate. However, the latter (STP-WE) also entails the highest cost and the lowest energy efficiency. The advantages and drawbacks of the hydrogen production schemes using solar energy are summarized in Table 7.5.

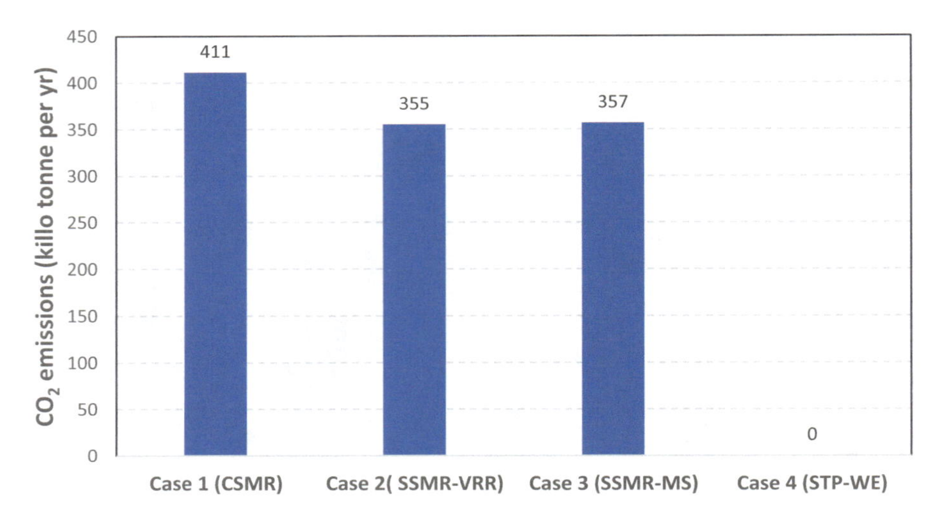

Fig. 7.11 CO_2 emission rate for hydrogen production schemes

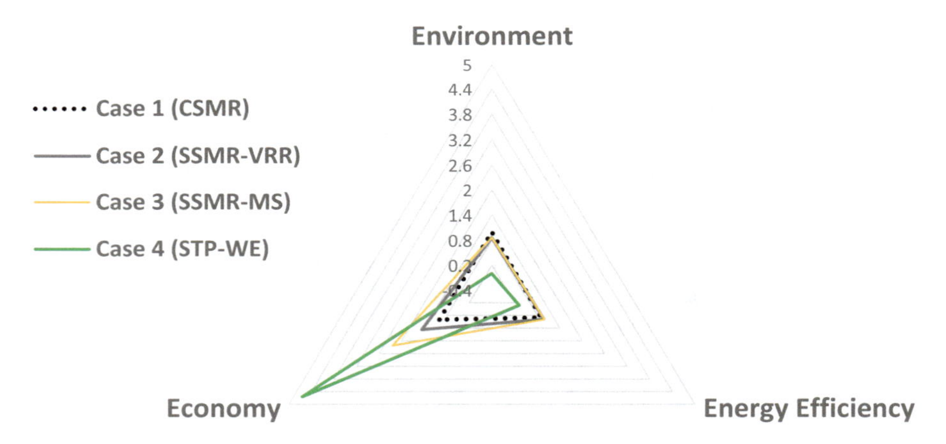

Fig. 7.12 Sustainability performance of hydrogen production schemes

7.6 Concluding Remarks

This chapter presents alternative schemes to integrate renewable energy with the oil and gas industry specifically for crude oil upgrading processes by using solar energy to produce the hydrogen required. Refineries today mainly produce hydrogen through steam methane reforming (SMR), but it emits a high level of greenhouse gases (CO_2). Our work proposes three hydrogen production schemes using solar energy, namely, solar steam methane reforming using volumetric receiver reactor (SSMR-VRR), solar steam methane reforming using molten salt as a heat carrier

Table 7.5 Pros and cons of solar-aided hydrogen production schemes for oil and gas applications

Scheme	Advantage	Disadvantage
CSMR	Moderate energy efficiency	High CO_2 emissions
SSMR-VRR	High energy efficiency Reduced natural gas consumption Reduced water consumption Lower CO_2 emissions Lowest LCHP	Limited operating time of volumetric receiver reactor High complexity due to the existence of two operating modes
SSMR-MS	High energy efficiency (vs. CSMR) Reduced natural gas consumption Reduced water consumption Low CO_2 emissions Thermal energy storage allows constant operation	Heat from concentrated solar can only be used to pre-reforming of natural gas
STP-WE	No CO_2 emissions, high-purity oxygen byproduct, clean and sustainable	Low energy efficiency; large land area; expensive equipment; highest LCHP

(SSMR-MS), and solar thermal power generation coupled with water electrolyzer (STP-WE). We compare their conceptual design using steady-state simulation with conventional SMR on criteria comprising technical (energy efficiency), economics (levelized cost), and environment.

Economic analysis of the schemes suggests that at current development stage, thermochemical processes (as in SSMR-VRR and SSMR-MS) hold greater potential to produce hydrogen required for a crude oil upgrader whereby SSMR-VRR gives a lowest LCHP. On the other hand, LCHP of electrochemical processes (as in STP-WE) is higher due to its capital investment cost for renewable energy–related equipment. In terms of environmental performance as compared to CSMR, SSMR-VRR is advantageous with lower CO_2 emissions and natural gas consumption, which is like SSMR-MS that shows a highest energy efficiency besides reduced CO_2 emission as well as natural gas and electricity consumption. Nonetheless, it is arguable that as renewables technology advances enough to reduce the associated capital cost, STP-WE process can be a more promising option for hydrogen production, because it is clean and sustainable as evidenced by a lowest greenhouse gas emission.

References

1. S. Koumi Ngoh, D. Njomo, An overview of hydrogen gas production from solar energy. Renew. Sust. Energ. Rev. **16**, 6782–6792 (2012)
2. OPEC, *World Oil Outlook 2045* (Organization of the Petroleum Exporting Countries, Vienna, 2020)
3. M. Absi Halabi et al., Application of solar energy in the oil industry – Current status and future prospects. Renew. Sust. Energ. Rev. **43**, 296–314 (2015)
4. Petroleum HPV Testing Group, *Crude Oil Category* (American Petroleum Institute, 2011)

5. M. Arriaga et al., Renewable energy alternatives for remote communities in Northern Ontario, Canada. IEEE Trans. Sustain. Energy **4**, 661–670 (2013)
6. T.M. Razykov et al., Solar photovoltaic electricity: Current status and future prospects. Sol. Energy **85**, 1580–1608 (2011)
7. F. Dinçer, The analysis on photovoltaic electricity generation status, potential and policies of the leading countries in solar energy. Renew. Sust. Energ. Rev. **15**, 713–720 (2011)
8. Y. Tian, C.Y. Zhao, A review of solar collectors and thermal energy storage in solar thermal applications. Appl. Energy **104**, 538–553 (2013)
9. D. Mills, Advances in solar thermal electricity technology. Sol. Energy **76**, 19–31 (2004)
10. L. Burnham et al. (eds.), *Renewable Energy: Sources for Fuels and Electricity* (Island Press, 1992)
11. F. Sayedin et al., Optimal design and operation of a photovoltaic–electrolyser system using particle swarm optimisation. Int. J. Sustain. Energy **35**, 566–582 (2016)
12. E. Wilhelm, M. Fowler, A technical and economic review of solar hydrogen production technologies. Bull. Sci. Technol. Soc. **26**, 278–287 (2006)
13. R.F. Meyer et al., *Heavy Oil and Natural Bitumen Resources in Geological Basins of the World* (U.S. Geological Survey, Reston, 2007)
14. J.G. Speight, Chapter 6 – Upgrading during recovery, in *Heavy Oil Production Processes*, ed. by J.G. Speight, (Gulf Professional Publishing, Boston, 2013), pp. 131–148
15. International Energy Agency (IEA), *Technology Roadmap – Hydrogen and Fuel Cells* (Paris, 2015)
16. Z. Li et al., Photoelectrochemical cells for solar hydrogen production: Current state of promising photoelectrodes, methods to improve their properties, and outlook. Energy Environ. Sci. **6**, 347–370 (2013)
17. A. Steinfeld, Solar thermochemical production of hydrogen – A review. Sol. Energy **78**, 603–615 (2005)
18. D. Yadav, R. Banerjee, A review of solar thermochemical processes. Renew. Sust. Energ. Rev. **54**, 497–532 (2016)
19. S. Möller et al., Hydrogen production by solar reforming of natural gas: A comparison study of two possible process configurations. J. Solar Energy Eng. **128**, 16–23 (2005)
20. A. Maroufmashat et al., An imperialist competitive algorithm approach for multi-objective optimization of direct coupling photovoltaic-electrolyzer systems. Int. J. Hydrog. Energy **39**, 18743–18757 (2014)
21. C. Agrafiotis et al., Solar thermal reforming of methane feedstocks for hydrogen and syngas production – A review. Renew. Sust. Energ. Rev. **29**, 656–682 (2014)
22. H.I. Villafán-Vidales et al., An overview of the solar thermochemical processes for hydrogen and syngas production: Reactors, and facilities. Renew. Sust. Energ. Rev. **75**, 894–908 (2017)
23. S.A.M. Said et al., A review on solar reforming systems. Renew. Sust. Energ. Rev. **59**, 149–159 (2016)
24. National Renewable Energy Laboratory (NREL), *System Advisor Model (SAM)* (2020, December 20), https://sam.nrel.gov/
25. Aspen Technology (AspenTech), *Design and Optimize Chemical Processes with Aspen Plus* (2020, October 22), http://www.aspentech.com/products/engineering/aspen-plus/
26. S. Walker et al., Implementing power-to-gas to provide green hydrogen to a bitumen upgrader. Int. J. Energy Res. **40**, 1925–1934 (2016)
27. C. Likkasit et al., Integration of renewable energy into oil and gas industries: Solar-aided hydrogen production, presented at the International Conference on Industrial Engineering and Operations Management (IEOM), Detroit, MI, USA (2016)
28. C. Likkasit et al., Solar-aided hydrogen production methods for the integration of renewable energies into oil & gas industries. Energy Convers. Manag. **168**, 395–406 (2018)
29. W.D. Seider et al., *Product and Process Design Principles*, 3rd edn. (Wiley, 2010)

30. T. Ramsden, *Current Central Hydrogen Production from Grid Electrolysis version 2.1.2* (2019), https://www.nrel.gov/hydrogen/assets/docs/current-central-grid-electrolysis-v2-1-2.xls. Accessed 11 July 2019
31. A. Maroufmashat, M. Fowler, Transition of future energy system infrastructure; through power-to-gas pathways. Energies **10**, 1089 (2017)
32. F. He, F. Li, Hydrogen production from methane and solar energy – Process evaluations and comparison studies. Int. J. Hydrog. Energy **39**, 18092–18102 (2014)
33. A. Giaconia et al., Solar steam reforming of natural gas for hydrogen production using molten salt heat carriers. AICHE J. **54**, 1932–1944 (2008)

Chapter 8
A Review on CO_2 Monitoring Satellites

Steve Houang, Andres Espitia, Shawn Pang, Joshua Cox, Ali Ahmadian, and Ali Elkamel

8.1 Introduction

Cities have a significant impact on climate governance. Although urban areas comprise just 2% of the earth's surface, they account for more than 70% of worldwide carbon emissions [1]. This ratio is growing as urbanization and energy use both increase [1]. For a better understanding of carbon emissions on a global scale and their influence on climate change, the evaluation of urban carbon emissions is therefore crucial.

Many previous studies for the evaluation of urban carbon emissions have focused on data based on carbon emission inventories. These data are collected from emission sectors such as electricity generation, industries, transportation, agriculture, and household and office use [2].

Currently, urban CO_2 emission inventories have been gathered in developed cities. These reports are usually prepared with different methods, so their analytical comparison is challenging [3].

Although estimating carbon emissions based on emission inventories statistically enhances our knowledge of urban carbon emissions, due to inaccuracies in the data or the data analysis method [4], it sometimes underestimates the real emissions,

S. Houang · A. Espitia · S. Pang · J. Cox
Department of Chemical Engineering, University of Waterloo, Waterloo, ON, Canada

A. Ahmadian (✉)
Department of Electrical Engineering, University of Bonab, Bonab, Iran

Department of Chemical Engineering, University of Waterloo, Waterloo, ON, Canada

A. Elkamel
Department of Chemical Engineering, University of Waterloo, Waterloo, ON, Canada

Department of Chemical Engineering, Khalifa University, Abu Dhabi, United Arab Emirates
e-mail: aelkamel@uwaterloo.ca

A. Ahmadian et al. (eds.), *Carbon Capture, Utilization, and Storage Technologies*, Green Energy and Technology, https://doi.org/10.1007/978-3-031-46590-1_8

particularly at the urban scale [5]. Several studies have employed ground-based CO_2 measurements to decrease uncertainty in emission inventory–based approaches [6]. The spatial and geographical constraint of terrestrial CO_2 observations is one of the most significant issues with these measurements. In other words, in certain regions of the globe, such as the Middle East, there are no CO_2 measuring stations on the ground. In addition, the high costs of station building and ground measurements limit their widespread usage, particularly in less developed regions [7].

Since data from inventory analysis of carbon emission may overestimate or underestimate actual carbon emissions [8], the IPCC guidelines recommend that studies using carbon emission inventory data should validate and confirm their results using ground-based CO_2 concentration data [9].

In order to implement measures to reduce CO_2 emissions and, subsequently, to reduce climate change on the metropolitan level, there is a need for more accurate and quantitative information on CO_2 emissions. Satellite data can provide an independent, easy, and accessible method to determine the amount of urban-level CO_2 emissions.

Many studies show the compatibility of XCO_2 satellite observation results with carbon emission inventory estimates and modeling [10–12]. As an example, the ability to measure CO_2 emissions using the SCIAMACHY meter resulted in the identification of XCO_2 values in urban areas as anthropogenic CO_2 sources [13]. Also, the use of GOSAT satellite observations to monitor urban CO_2 emissions compared to the terrestrial monitoring network (TCCON) has been able to detect changes in CO_2 emissions at a 95% confidence level [14]. OCO-2 satellite has also been able to detect XCO_2 values and the enhancement of anthropogenic XCO_2 in urban, suburban, and rural areas with high accuracy [10, 14, 16]. Since 2019, the OCO-3 satellite with SAM observation mode has shown high ability to monitor CO_2 emissions, especially in metropolises. These results show that the observation of CO_2 concentration by satellites, together with ground measurements, provides reliable data sources for estimating CO_2 emissions caused by human activities and also confirming carbon emission inventories [17].

The enhancement of CO_2 concentration in large cities such as Beijing, Tianjin, and Hebei regions in China, Los Angeles metropolitan area in America, Seoul metropolitan area in South Korea, and Mumbai in India has been estimated to be more than 3 ppm based on satellite remote sensing observations [13, 15, 18–21]. Overall, compared to the global annual average of XCO_2, which is estimated at 414.71 ppm in 2021, the increase in CO_2 concentration in cities is between 0 and 6 ppm and generally less than 3 ppm [22], which is a small contribution in measuring the total column of CO_2 concentration by satellite meters. Therefore, the accuracy of CO_2 emission estimates with the help of satellite observations in cities is highly sensitive to environmental conditions such as wind and atmospheric CO_2 transport, which should be considered before using these data to monitor urban CO_2 emission.

In general, increased CO_2 levels in the point source emission sites such as urban areas, power plants, and volcanoes can be detected by removing the background concentration effect [16]. That is, the increase in CO_2 concentration can be obtained by determining the difference between the observed values at the emission source and the values in a clean area (as background) near the source. These values indicate the real contribution of cities in increasing the concentration of CO_2 in the atmosphere and are compatible with the direct emission of carbon from urban areas [16, 22].

In recent years, many satellites and sensors have been placed in orbit to measure greenhouse gases in the atmosphere (especially CO_2). Their recorded observations have a high variety in terms of spatial and temporal resolution, accuracy, collection history, etc. Comparing the spatial resolution of the satellites discussed in Sect. 8.3 shows that Aura, GOSAT, OCO-2, TANSAT, GOSAT-2, and OCO-3 satellites perform better in estimating CO_2 emissions at urban scales due to their lower spatial resolution.

To detect the concentration of CO_2 on a regional and urban scale, the concentration in the lower layers of the atmosphere and areas close to the earth's surface is very important. The Aura satellite measures the concentration of CO_2 in the middle troposphere [23], but other satellites estimate the concentration of the CO_2 column in the near earth's surface [24–27]. Therefore, the Aura satellite is not used to investigate anthropogenic CO_2 emissions in cities.

Both GOSAT and GOSAT-2 satellites have a spatial resolution of 10×10 km. GOSAT has an accuracy of 4 ppm [26]. With the improvements and progresses made on GOSAT-2, its precision has been greatly improved compared to GOSAT. Although GOSAT has lower precision, it has a longer time history and its measurements has been available since 2009 [26, 28].

OCO-2 and OCO-3 satellites have better spatial resolution than other satellites. These two satellites have recorded atmospheric CO_2 concentration since 2014 and 2019, respectively, and are currently considered among the best satellites for measuring CO_2 emissions on small scales (regional, urban, and local). The accuracy of both satellites is less than 1 ppm [24, 27].

According to the comparisons, it can be stated that OCO-2 and OCO-3 are useful in detecting small-scale CO_2 emissions due to their proper spatial resolution and high accuracy. On the other hand, GOSAT and GOSAT-2 satellites have a shorter acquisition time cycle than OCO-2 (3 and 6 days, respectively) [26, 28].

More accurate satellites such as GOSAT-2 and satellites with higher spatial resolution such as OCO-2 and OCO-3 have been launched in the last decade and do not have long-term data recording history; however, GOSAT gives older data than these satellites (since 2009) to study the history of CO_2 concentration.

Observation and collection of GOSAT data is a grid of widely spaced points with limited observations in certain areas [29]. As a result, in every time passing through the emission areas such as big cities, due to the scattering of the collection points, it covers only small parts of them.

GOSAT-2 is more focused on observing point sources in the target mode than its predecessor (GOSAT); however due to the scattering of collected data, it cannot entirely cover urban emission zones. Compared to GOSATs, the number of observations of the total CO_2 column concentration by the OCO-2 satellite has grown dramatically, and this spacecraft gathers around one million data every day [30]. This satellite is equipped with high-precision spectrometers intended to detect

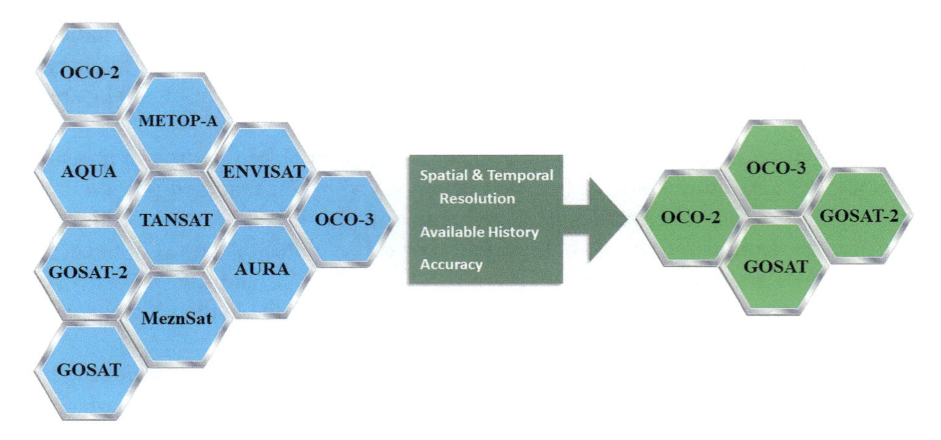

Fig. 8.1 Choosing the right satellite for urban-scale anthropogenic emission monitoring. GOSAT, OCO-2, GOSAT-2, and OCO-3 are recommended for studying urban emissions due to their time history, appropriate spatial and temporal resolution, and high accuracy

small-scale CO_2 variations. The harvesting strip on OCO-2 is about 10 km wide [30]. Consequently, compared to the GOSAT and GOSAT-2 satellites, it can cover a greater urban region. The OCO-3 satellite can measure more data from significant emission locations (such as cities, power plants, and volcanoes) due to its greater spatial coverage and shorter time cycle [22]. OCO-3 incorporates a method of observation intended to assess human-caused emissions. This mode, referred to as "Instantaneous Imager Area Map (SAM)," produces dense data from XCO_2 covering an area of roughly 80×80 km^2. This method of data gathering covers the majority of urban emission locations. The number of OCO-3 observations in the SAM mode, which is multi-band, is about three times more than the number of OCO-2 observations (observation in a single band) [22].

- Therefore, according to the strengths and weaknesses of each of these satellites, one can conclude that the simultaneous use of the data of these four satellites has several major advantages, which include obtaining historical data since 2009 using the GOSAT satellite.
- Study of shorter time cycles by GOSAT, GOSAT-2, and OCO-3 satellites.
- Delivering higher accuracy by using OCO-2, OCO-3, and GOSAT-2.
- Smaller spatial resolution by OCO-2 and OCO-3.
- Dense spatial coverage of urban emission sources by OCO-3.

Therefore, the simultaneous use of these four satellites is suggested in order to acquire an accurate estimate for emissions with more detailed spatial resolution and better precision (Fig. 8.1).

8.2 GOSAT Satellite

The GOSAT satellite was launched on January 23, 2009 in cooperation with the Japan Aerospace Exploration Agency (JAXA), the Ministry of Environment (MOE), and the National Institute of Environmental Research (NIES) to study the transport mechanisms of greenhouse gases such as CO_2 and CH4 [26]. In this project, JAXA is responsible for developing the satellite and its instruments, launching the satellite, and operating it (including information acquisition). NIES is responsible for data extraction and analysis. MOE is responsible for project development and financing [29]. The overall goal of launching this satellite was to help manage the environment by estimating the amount of greenhouse gases, especially CO_2, and reducing these types of gases [29].

GOSAT is the first satellite to detect CO_2 and CH4 greenhouse gas concentrations using the short-wave infrared (SWIR) band. A month after the launch of this satellite in January 2009, the operation to evaluate its performance began. The data measured by GOSAT have different levels [31].

Initially, the level one results were reported after compatibility tests with XCO_2 and XCH_4 ground measurements [31] were conducted. In February 2010, JAXA conducted early validations and released the results of level two XCO_2 and XCH_4 surface data to the public. The data was then elevated to level three. The GOSAT satellite commenced its active phase in October 2011. In the summer of 2012, this satellite officially started providing level four data [31]. NIES delivers high-resolution data from the GOSAT satellite with an inaccuracy of less than 1% every 3 days [32].

The orbit of the GOSAT satellite is the sun-synchronous, which moves at a distance of 666 km from the earth's surface, under an inclination angle of 98°, and completes each rotation in approximately 100 min. This satellite crosses the equator at 13:00 $\pm$ 15 h [31]. The data of this satellite can be accessed through the database of the European Center for Medium-Term Climate Predictions (ECMWF) / European Copernicus Climate Change Service (available at: https://cds.climat.copernicus.eu/cdsapp).

Due to the fact that the presence of fine aerosols reduces the accuracy of XCO_2 estimate, the GOSAT employs the spectrometer of Cloud and Aerosol Imager (TANSO-CAI) [26] to mitigate this impact and present the XCO_2 with 1% accuracy [31].

8.2.1 TANSO-FTS

The GOSAT satellite is equipped with the Thermal and Near-Infrared Instrument for Carbon Observation (TANSO), which uses a Fourier Transform Spectrometer (FTS). This sensor has a wide spectrum coverage from VIS to TIR (Table 8.1) [26]. Bands 1, 2, and 3 of this meter are only able to record information during the

Table 8.1 Spectral characteristics of the TANSO-FTS meter [35]

Spectral band no	1	2	3	4
Spectral range	VIS	SWIR	SWIR	MWIR/TIR
Coverage (μm)	0.758–0.775	1.56–1.73	1.92–2.09	5.5–14.3
Target gases	O_3	CH_4, CO_2	CO_2	CH_4, CO_2, O_3
Spectral resolution (cm^{-1})	0.5	0.2	0.2	0.2

Table 8.2 Spectral characteristics of TANSO-CAI measurement bands. This meter can detect fine particles and cloud cover and increase the

Spectral band no	1	2	3	4
Center wavelength (μm)	0.380	0.674	0.870	1.6

day. However, band 4 (MWIR/TIR spectral range) does not require sunlight to collect data [31]. TANSO-FTS meter includes four spectral bands that measure O_2, O_3, CO_2, and CH4 gases (Table 8.1).

8.2.2 TANSO-CAI

In addition to the TANSO-FTS instrument, the GOSAT satellite is equipped with the Thermal and Near-Infrared Instrument for Carbon Observation (TANSO), which used a recorder tool for clouds and aerosols. TANSO-CAI is equipped with an electronic imager that makes very detailed observations of high-altitude cirrus clouds and the dispersion of suspended particles (even fine dust in near-earth regions) in the ultraviolet (UV), visible and short-wave infrared (SWIR) spectral ranges [31]. This meter is used to correct the interference of clouds and the effect of fine dust on the values of XCO_2 and XCH_4 [26] (Table 8.2). This imager has continuous spatial coverage, wider field of view, and higher spatial resolution compared to FTS [26] to detect micro-dust and cloud cover (Fig. 8.2).

8.2.3 GOSAT Observation Mode

The mechanism for determining the data collection points by the TANSO-FTS meter is a two-axis pointing function for cross tracks (CT) and along tracks (AT), which allows accurate observation of the ground [29]. This sensor covers a scanning angle of ±35° from the nadir direction (Fig. 8.3) [29] to observe any target on Earth's surface in a three-day repetition cycle. TANSO-FTS performs CO_2 measurement in several different modes, which include:

Grid point observation mode: In this mode, data collection is done as a grid with two axes along the length (AT) and width (CT) of the satellite movement.

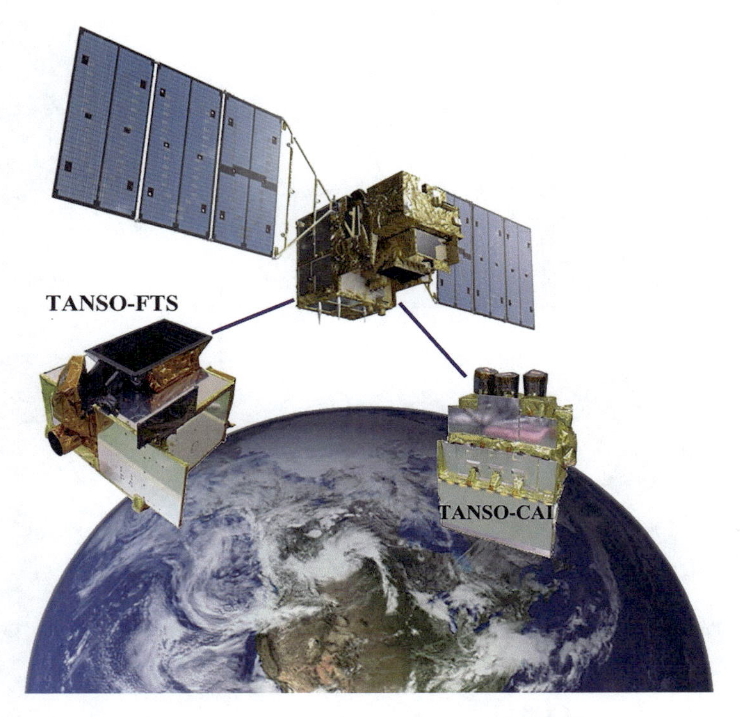

Fig. 8.2 GOSAT satellite and the position of two sensors TANSO-CAI and TANSO-FTS

TANSO-FTS can collect observations up to an angle of ±35° in the CT direction and ± 20° in the AT direction [29].

The field of view of TANSO-FTS includes 1 to 9 points in the transverse direction in each sampling strip [33]. The harvesting mechanism in the CT direction can also include 1, 3, 5, 7, and 9 harvest points. The distance of these points varies from 790 (in single point mode) to 88 km (in nine-point mode) [33]. The instant field of view at each sampling point is 10.5 km [29]. Figure 8.4 shows the observation geometry in the case where the field of view of the sensor is 750 km and includes 5 sampling points in the transverse direction.

TANSO-CAI also has a wide range of observations with a width of 1000 km and a viewing angle of ±35° in the CT direction [31].

Sun-glint observation mode: This type of observation usually takes place over the ocean where the surface reflection is small. In the mode of observing light reflection, the sensor observes and measures this reflection in areas where the amount of sunlight reflection is high [34].

Since the viewing angle range in the longitudinal direction (AT) is ±20°, this observation mode is limited to low and middle latitudes [34] (Fig. 8.5).

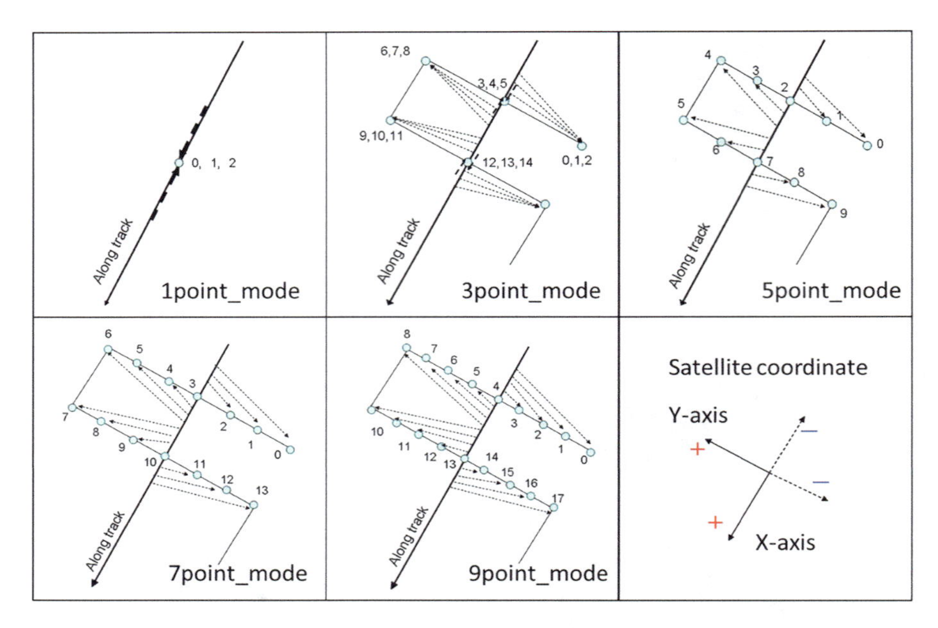

Fig. 8.3 Patterns of observation by the TANSO-FTS sensor in different fields of view [34]

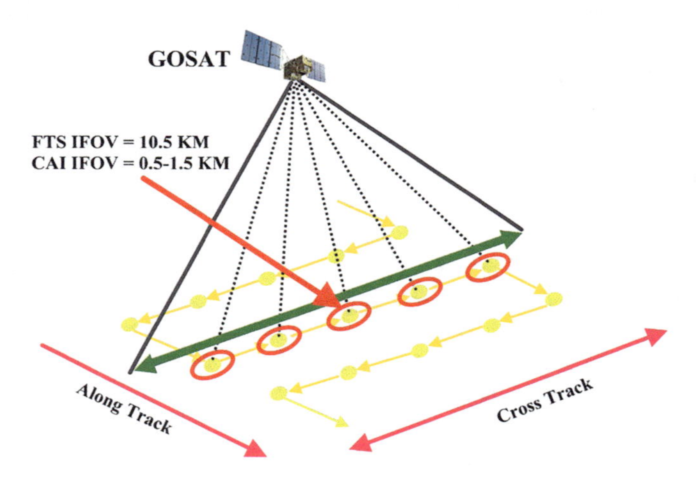

Fig. 8.4 Observation geometry by TANSO-FTS. The field of view of the sensor is 750 km and includes five sampling points in the transverse direction [33]

Target mode observation: TANSO-FTS meter can view the targets between ±20° in the AT direction and ± 35° in the CT direction by adjusting the set of view angles. This type of observation is for validation and calibration (indirect) as well as observation of large emission sources such as large cities, active volcanoes, oil and gas fields, and landfills (Fig. 8.6) [35].

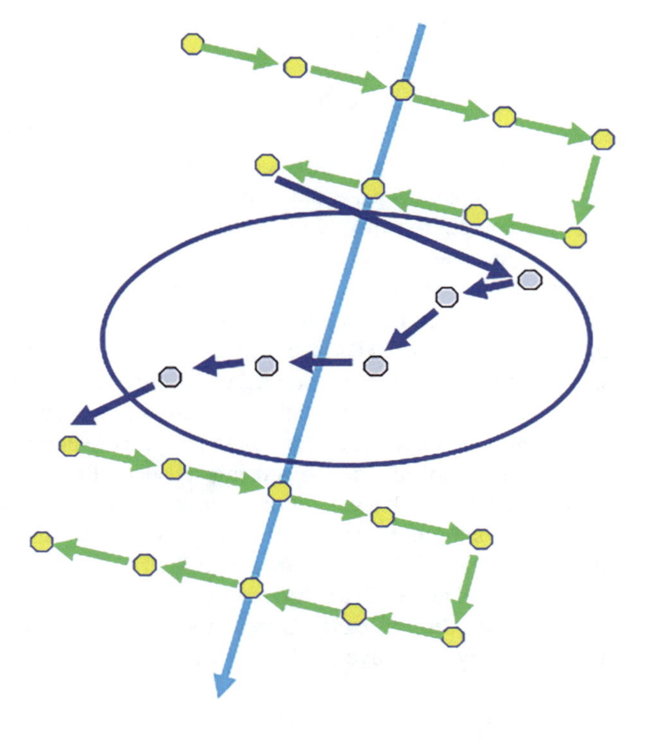

Fig. 8.5 Geometry (Geometry) observation of TANSO-FTS in the target mode. This meter adjusts the observation angles so that it observes the targets between the angles of $\pm 20°$ in the AT direction and $\pm$ 35° in the CT direction [34]

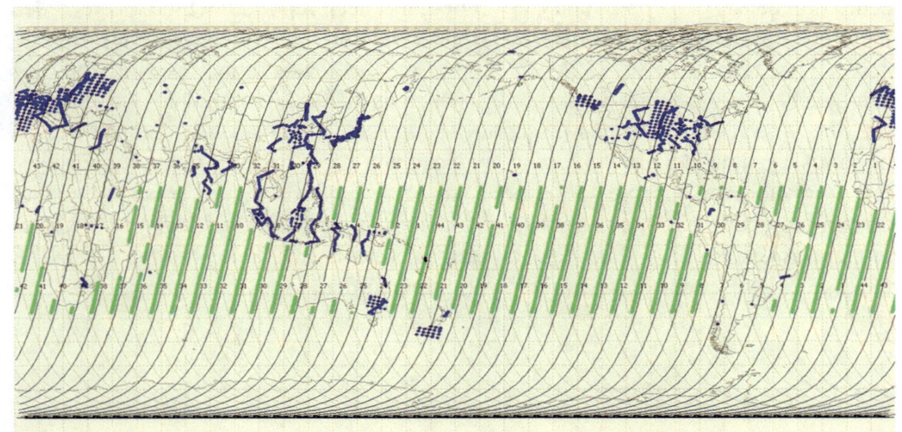

Fig. 8.6 An example of the TANSO-FTS observation pattern. The green points show the observations in the sunlight reflection mode and the blue points display the observations in the target mode [34]

8.3 GOSAT-2 Satellite

After launching the GOSAT satellite on January 23, 2009 in collaboration with the Japan Aerospace Exploration Agency (JAXA), the Ministry of the Environment (MOE), and the National Institute of Environmental Research (NIES), these three organizations launched a second satellite designated GOSAT-2. It was sent into orbit after the completion of earlier satellite operations [36]. The GOSAT satellite has successfully completed the validation and calibration procedures, resulting in enhancements such as the detection of greenhouse gases and more precise measurements of their concentration. GOSAT-2 was launched in October 2018 with upgraded FTS and CAI sensors [30] in order to enhance the precision of XCO_2 measurements. This was accomplished by enhancing the measurement accuracy and achieving an appropriate spatial resolution. This satellite was managed by the aforementioned three organizations in an effort to conduct more precise and extensive investigations and analyses using sensors with improved performance in order to monitor climate change and ultimately decrease greenhouse gas emissions [28].

A few months after the launch of this satellite, first the TANSO-CAI-2 sensor and then the TANSO-FTS-2 sensor started their activities, and as a result, the correct operation of all GOSAT-2 components was confirmed. At the beginning of February 2019, this satellite officially entered the operational stage. Level 2 data obtained from XCO_2 by the above satellite was released to the public 1 year after launch, in October 2019 [36].

The accuracy of observations of the GOSAT-2 satellite for measuring the concentration of CO_2, 0.5 ppm, and CH_4, 5 ppb, has been reported, which has recorded more accurate observations than the GOSAT satellite (4 ppm and 34 ppb for CO_2 and CH_4, respectively) [30]. In addition, the observation period for this sun-synchronous satellite is 1 month, which is faster compared to the GOSAT satellite, which was 3 months [30]. Other successes of this satellite compared to its previous version include increasing the accuracy of estimating natural emissions, focusing more to observe target-point sources, monitoring the concentration of carbon monoxide (CO), and the use of CO and nitrogen dioxide (NO_2) to calculate human emissions [30]. The data of this satellite is available through the NIES GOSAT-2 Project database (available at: https://prdct.gosat-2.nies.go.jp).

8.3.1 TANSO-FTS-2

This sensor is one of the important components in the structure of the GOSAT-2 satellite, which is responsible for checking the density of greenhouse gases in the atmosphere. Compared to its previous version, TANSO-FTS-2 meter has improved features such as more accurate calibration and a faster and more intelligent pointing system. Moreover, this sensor has made it possible to measure greenhouse gases in the atmosphere in cloudy conditions [28, 36]. This tool has been able to identify the

Table 8.3 Comparison of specifications of the GOSAT and GOSAT-2 satellites [31, 37]

Satellite mission	GOSAT	GOSAT-2
Launch date	23 January 2009	29 October 2018
Mission type	Earth observation	
Agency	MOE (Japan), JAXA, NIES (Japan)	
Measurement domain	Atmosphere, land	
Measurement category	Aerosols, vegetation, cloud type, amount and cloud top temperature, atmospheric temperature fields, ozone, trace gases (excluding ozone)	
Gases measured	CO_2, CH_4, O_3, H_2O	CO_2, CH_4, O_3, H_2O, CO, NO_2
Instruments	TANSO-FTS, TANSO-CAI	TANSO-CAI-2, TANSO-FTS-2
Altitude (km)	666	613
Repeat cycle (day)	3	6
Spatial resolution (km)	~ 10 × 10	~ 10 × 10
Relative accuracy	4 ppm for CO_2 and 34 ppb for CH_4	0.5 ppm for CO_2 and 5 ppb for CH_4
CO_2 and CH_4 sensitivity	Total column including near surface	
Power (EOL) (KW)	3.8	5

Table 8.4 Spectral specifications of TANSO-FTS-2 meter [28, 37]

Spectral bands no	Observation bands (mμ)	Spectral resolution (cm^{-1})	Mission purpose
1	0.755–0.772	0.2	Total column O_2
2	1.563–1.695	0.2	Total column CO_2 and CH_4
3	1.923–2.381	0.2	Total column CO_2, moisture, and CO
4	5.56–8.45	0.2	CH_4 and moisture
5	8.45–14.3	0.2	Temperature profile, CO_2

sources of greenhouse gas emissions with appropriate accuracy, which this issue has attracted the attention of many experts in the field of climate change [30]. This meter has 5 spectral bands ranging from 0.755 to 14.3 [30] (Table 8.3).

Observation modes in GOSAT-2 are largely similar to GOSAT, with the difference that TANSO-FTS-2 is able to observe targets up to an angle of ±40° in the AT direction and ± 35° in the CT direction [36]. The viewing angle of this sensor in the AT direction has increased from 20° in the GOSAT satellite to 40°, which increases the visible area over the ocean (measurement in the sunniest state) [37]. This capability provides the possibility of increasing the number of observation points on the ocean and contributing to global observation, including over the ocean [37]. Table 8.4 represents the spectral specifications of TANSO-FTS-2 meter.

This sensor is able to detect cloudless areas and place the scanner in the field of view (FOV). In this way, the number of cloud-free measurements without clouds increases. This capability is called intelligent pointing (IP). IP checks the position of clouds during the rotation time in the orbit and changes the field of view in the appropriate direction to obtain high-quality and cloud-free observations [36].

8.3.2 TANSO-CAI-2

The observation bands of this sensor capture Earth's surface with an angle of ±20° forward and backward (Fig. 8.7). This meter is used for imaging and checking the optical properties of airborne particles and clouds, as well as monitoring the state of urban and atmospheric air pollution. One of the important tasks of this meter is to separate clouds in different directions [38, 39].

Two examples of the equipment used in this satellite in the cloud separation system include:

- Using an RGB camera (on-board camera (CAM) is an RGB camera). This camera checks the presence or absence of clouds in the sampling strip of the TANSO-FTS-2 meter, and if there are clouds at the sampling point, it automatically changes the spotting angle of the meter. With this system, the probability of obtaining cloud-free measurements is increased and, as a result, the measurement accuracy also increases [36].
- The TANSO-CAI-2 sensor has a near-infrared band (0.87 μm) and a short-wave infrared band (1.60 μm), which are effective in detecting clouds [36].

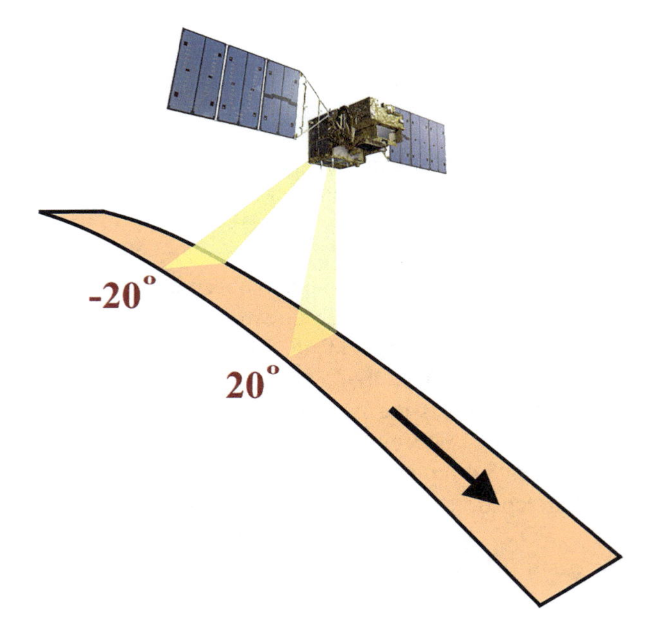

Fig. 8.7 TANSO-CAI-2 imaging range. This sensor captures Earth's surface at an angle of ±20° forward and backward [39]

8.4 OCO-2

The National Aeronautics and Space Administration (NASA) in the United States has dedicated the OCO-2 satellite to the study of carbon dioxide in the atmosphere. This satellite has the ability to provide a complete picture of human and natural carbon dioxide sources and sinks of CO_2 [30].

The OCO-2 satellite, which was launched on July 2, 2014, is actually a copy of the OCO satellite, which was decommissioned in 2009 due to technical problems. This satellite is on sun-synchronous orbit A-Train, at an altitude of 705 km from Earth's surface, moves with an inclination angle of 2.98° and crosses the equator at around 1:30 pm local time. Its rotation period is equal to 98.8 min and the measurement repetition period is 16 days [30].

The satellite has high-precision spectrometers designed to detect small fluctuations in XCO_2. OCO-2 measures the radiances reflected from Earth's surface in spectral bands near 0.765 μm, 1.61 μm, and 2.06 μm. OCO-2 collects data along eight parallel acquisition strips whose total width is less than or equal to 10 km. OCO-2 observations have a smaller footprint from 1.29 to 2.25 km [30].

Clouds are known to be a source of error in remote sensing measurements. The advantage of measuring with appropriate spatial resolution in the OCO-2 satellite observations increases the number of cloud-free measurements and therefore allows retrieval with fewer errors [40]. This satellite has dramatically increased the number of atmospheric CO_2 observations, collecting one million measurements per day when the satellite flies over the sunlit hemisphere [30]. The data of this satellite is available through the Goddard Earth Sciences Data & Information Services Center (available at: https://disc.gsfc.nasa.gov/datasets).

This satellite collects global XCO_2 measurements with high spatial and temporal resolution and high accuracy. With the help of these measurements, regional changes of CO_2 gas in Earth's climate system can be studied more accurately. OCO-2 is the first NASA satellite specifically designed to measure atmospheric CO_2 at regional scales. This satellite can even detect changes in CO_2 concentration from smaller and isolated areas such as individual cities. One of the most interesting findings of the OCO-2 satellite in this area was the observation of a strong signal of CO_2 concentration in the Middle East, which is not present in the emission inventories, indicating that the emission inventories may be incomplete in this region [6].

8.4.1 OCO-2 Observation Mode

The OCO-2 satellite collects data in three modes, nadir, glint, and target. Different modes optimize the sensitivity and accuracy of observations for specific applications (Fig. 8.8).

Nadir mode: The sensor directly observes the location of the ground under the satellite (local nadir). Rare observations provide the best spatial resolution, and

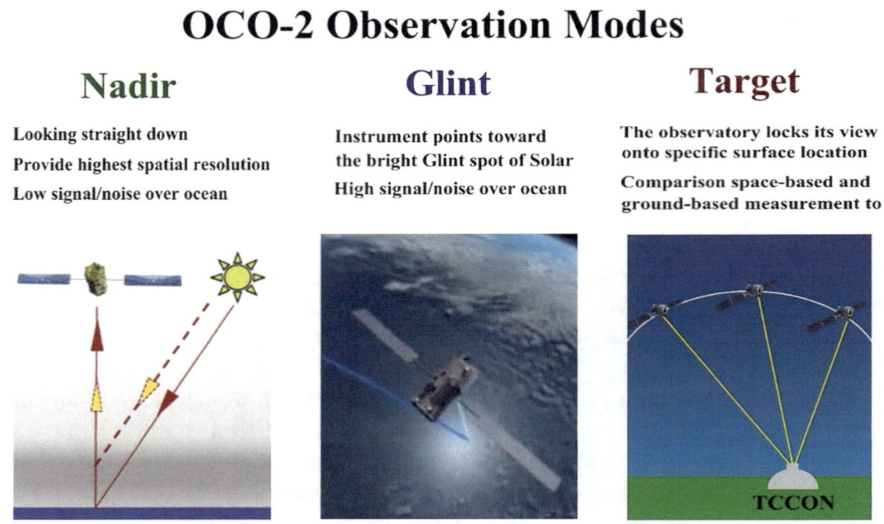

OCO-2 Observation Modes

Nadir

Looking straight down
Provide highest spatial resolution
Low signal/noise over ocean

Glint

Instrument points toward
the bright Glint spot of Solar
High signal/noise over ocean

Target

The observatory locks its view
onto specific surface location
Comparison space-based and
ground-based measurement to

TCCON

Fig. 8.8 Schematic view of OCO-2 satellite observation modes [41]

in regions that are partly cloudy or topographically rugged, rare observations provide more useful XCO_2 retrievals. In addition, these observations provide more useful data at high latitudes where sunlight is low [30].

Glint mode: The meter observes a location where sunlight is directly reflected from Earth's surface (a bright spot where solar radiation is specifically reflected from the surface). The glow mode enhances the meter's ability to obtain highly accurate measurements, especially in the ocean. The signal-to-noise ratio (SNR) of glow observations over dark surfaces and oceans is much higher than other regions, thus providing more accurate XCO_2 retrievals in these regions [30].

OCO-2 took its initial observations in 16-day time cycles and in only one nadir or glint mode, and after completing the cycle, the observation mode changed; so the whole land was harvested in both cases in approximately monthly time scales [30]. This approach covered oceans and continents on monthly time scales but produced 16-day long gaps in the ocean coverage in a rare state. In the glow mode, this method restricted the measurements in high latitudes. In early July 2015, the observational strategy between the glint and nadir states in alternating orbits was revised. This approach now continuously covers the bright hemisphere every day [30].

Target mode: The OCO-2 meter continuously observes it when passing over a specified range (usually a XCO_2 ground measurement site). This capability provides the ability to collect thousands of measurements at locations where alternative, accurate ground-based, and airborne instruments also measure atmospheric CO_2 [27, 30].

8.5 OCO-3

Following the success of the OCO-2 satellite in the field of XCO_2 data acquisition, a new version called the OCO-3 satellite was sent to the International Space Station (ISS) by NASA on May 4, 2019. This satellite was sent into space with the aim of observing information about the distribution of CO_2 on Earth, which is related to the growth of urban population and changing patterns of fossil fuel combustion [42]. Although most of the components of OCO-3 are similar to OCO-2, additional equipment has been added [42].

The satellite includes three high-resolution spectrometers that collect CO_2 measurements with the accuracy, resolution, and coverage needed to assess spatial and temporal variations of CO_2 over an annual cycle to enable the understanding of sources of CO_2 production and consumption at regional scales [42]. The two spectrometers of this satellite record the wavelengths in which the absorption of reflected radiation by CO_2 molecules is strong. A third spectrometer measures the wavelengths at which the absorption of reflected radiation by oxygen molecules is strongest. By combining the data from these three spectrometers, very precise measurements of XCO_2 are obtained by this satellite [43].

The OCO-3 satellite is capable of measuring and preparing maps of XCO_2 with high resolution. Combining the observations of this satellite with the observations of the OCO-2 satellite, the most accurate maps of the effects of humans and plants on the carbon cycle and thus on the climate can be drawn [16, 22]. The data of this satellite is available through the Goddard Earth Sciences Data & Information Services Center (available at: https://disc.gsfc.nasa.gov/datasets).

Unlike other CO_2 monitoring satellites, the OCO-3 satellite can scan large continuous areas that are hotbeds of CO_2 emissions (such as cities, power plants, and volcanoes). These measurements lead to dense and microscale spatial maps of XCO2 [22].

Recent studies have shown that CO_2 data collected by the OCO-2 satellite can be used to detect urban emissions [44–46]. However, the OCO-2 satellite data have problems in providing CO_2 concentration at urban and smaller scales. OCO-2 has a long repeat cycle of 16 days. In addition, the detection band of this satellite is narrow (10.3 km). This makes only a small part of the urban emissions data to be detected during a satellite pass (if the meteorological conditions are favorable) [22]. Moreover, OCO-2 moves in a sun-synchronous orbit, which means that all observations are limited to local afternoon. The OCO-3 satellite includes an observation mode specifically designed to measure anthropogenic emissions, overcoming some of the limitations mentioned above [22].

8.5.1 OCO-3 Observation Mode

The OCO-3 satellite collects the solar radiation spectrum in the illuminated hemisphere in four observation modes: nadir, glow, target, and SAM [22]. Although the three modes of OCO-3's nadir, glow, and target are similar to those of OCO-2, in OCO-3, it is possible to quickly switch between different observation modes. So the transition from the rare mode to the reflection mode for this satellite takes less than a minute [43].

Snapshot area map (SAM) mode: This new feature has made it possible for the first time to detect and map the XCO_2 differences at local scales. SAM's observation mode capability can measure CO_2 emissions from small emission sources (from power plants to large urban areas) in just 2 min, while it takes several days to measure the concentration of CO_2 in such areas by OCO-2 [31, 43]. In the SAM observation mode, OCO-3 collects data in closely spaced bands. This type of acquisition leads to the production of dense and micro-scale maps of XCO_2 that cover about 80×80 km^2 and are collected almost simultaneously, while other satellite instruments do not have this capability [22]. The number of observations of OCO-3 in the SAM mode is about three times more than the number of observations in the band-pass mode (OCO-2) [22].

During the observation of the target state, this satellite collects data in the form of sets of several swaths (usually 5 or 6 swaths), which overlap each other along the path. The overlap of these bands can cover an area of up to 20×80 km^2, which is a larger area compared to OCO-2 (20×20 km^2) [22].

Besides, OCO-3 has a larger footprint than OCO-2. The dimensions of its footprint in its nadir state are about 2.2 km in length and ≥ 1.6 km in width, which covers an area of 3.5 km^2, while the OCO-2 footprint covers an area of 3 km^2 [22].

Moreover, this satellite is able to sample megacities (one of the most important sources of greenhouse gas production and fossil fuel consumption), volcanoes, and other point emission sources and investigate the difference in CO_2 concentrations on a local scale for the first time from the space [22].

8.6 TCCON Validation

The first TCCON station was established in 2004. This station provided ground-based measurements of XCO_2 to help better understand the sources of production and consumption of CO_2 and CH_4 in Earth's atmosphere. After the construction of the first station, the TCCON network expanded in the following years. These stations provide important information on carbon-containing gases from many locations around the world [47] (Fig. 8.9).

TCCONs are a network of ground-based Fourier transform spectrometers that record the solar spectrum mainly in the near-infrared spectral region. This network

Fig. 8.9 TCCON ground stations including currently operational stations, expired stations, and stations to be launched in the future. This image was taken from the main page of the TCCON database [52]

can measure gases such as CO_2, CH_4, CO, N_2O, H_2O, HF, and HDO. These stations can detect small amounts of CO_2 gas in the atmospheric column at the location of the station and report the average concentration values in the entire CO_2 column [48, 49].

The main limitation of measurements at TCCON stations is that the spectrometer cannot record measurements when it is not sunny. For example, there are no measurements at night or under heavy cloud cover conditions [48]. In the absence of clouds, measurements are made at these stations approximately every 2 min. Data from each station provide information on carbon production and consumption sources at a regional scale. Additionally, by combining data from all stations, researchers can monitor carbon exchanges between the atmosphere, land, and ocean [47].

CO_2 observations from TCCON stations have shown that in the period of 2004–2014, the amount of CO_2 (XCO_2) recorded has increased by more than 20 ppm. This value has reached more than 400 ppm in the winter of 2014 in all stations of the Northern Hemisphere [47] and it increases on average about 2.2 ppm per year [50].

The Terrestrial Carbon Monitoring Network (TCCON) is commonly used to validate satellite XCO_2 measurements. To this end, satellites measure CO_2 in the vicinity of TCCON stations. By comparing the remotely sensed values with the values measured at TCCON stations, a complete assessment of the accuracy of CO_2 measurement by satellites is provided [51]. The GOSAT, GOSAT2, OCO-2, and OCO-3 satellites make many measurements while passing over the TCCON stations [74]. Therefore, most of the researches done by these satellites have been conducted at the metropolitan level, in cities or regions that have TCCON stations [27].

8.7 Benefits and Limitations

Remote sensing data is an effective data source for understanding the contribution of natural ecosystems and human activities to the increase of atmospheric XCO_2 at regional and global scales [53]. CO_2 monitoring satellites are different from each other in terms of specifications such as accuracy and collection history, spatial coverage, spatial and temporal resolution, etc. As mentioned, the GOSAT, GOSAT-2, OCO-2, and OCO-3 satellites are effective in detecting CO_2 emissions at regional and urban scales due to the measurement of the entire XCO_2 column, including areas close to Earth's surface [24, 37, 54, 55]. Although the acquisition history of GOSAT is longer than the other three satellites (from 2009 to present), it is less accurate (ppm 4) [30]. In terms of spatial resolution, both GOSAT and GOSAT-2 have a spatial resolution of 10×10 km [36], while OCO-2 and OCO-3 show a much better spatial resolution than these two satellites and therefore in detecting CO_2 emissions at urban and local scales provide more accurate measurements [43, 55].

The GOSAT and GOSAT-2 satellites have a shorter acquisition time cycle than the OCO-2 (3 and 6 days, respectively) [26, 28]. OCO-3 also has a variable harvesting cycle due to being equipped with dynamic observation-mode and flexible harvesting system [24].

In terms of data spatial dispersion, the GOSAT observations are a network of scattered and limited points with large distances [29]. As a result, each time passing through the emission areas, it covers only small parts of the area in different time frames (day, week, month, and year) (Fig. 8.10).

Despite the modifications made in GOSAT-2, the number of XCO_2 observations has increased compared to the previous version, but still due to the scattering of the data collected by GOSAT-2, urban and local emission areas are not completely covered by this satellite. The number of XCO_2 observations in the OCO-2 satellite has increased significantly compared to GOSATs [30]. Therefore, the data collected by OCO-2 is less scattered and covers much more parts of an urban area. Like OCO-2, the OCO-3 satellite records a large number of observations. Furthermore, due to the SAM observation mode, it can identify emission areas at small urban and local scales [22] (Fig. 8.10). The SAM observation mode is specifically designed to measure anthropogenic emissions. This capability allows OCO-3 to sample the entire surface of a large city in a short time frame and provide a large amount of XCO_2 data [22].

In general, despite the history of long-term acquisition and short time cycle, GOSAT is less useful in small-scale emission studies due to its low accuracy, small spatial resolution, and scattering of observations.

Although GOSAT-2 has a very high accuracy, the short acquisition history (since 2019), scattering of observations, and low spatial resolution have caused limitations in the detection of regional and local emissions.

OCO-2 has advantages such as high accuracy, good spatial resolution, spatial coverage, and a relatively large number of observations and has a long-time cycle

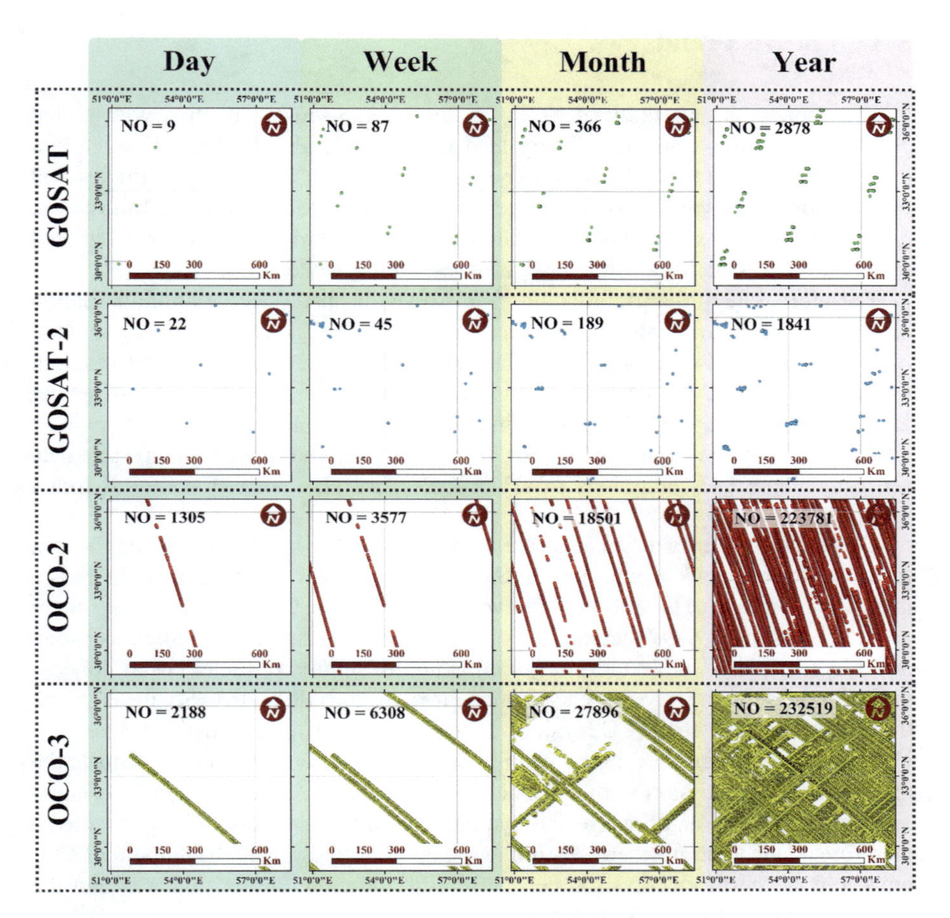

Fig. 8.10 Spatial coverage of GOSAT satellites; (**a**) GOSAT, (**b**) GOSAT-2, (**c**) OCO-2, and (**d**) OCO-3 in daily, weekly, monthly, and yearly time scales. OCO-3 and OCO-2 have the highest harvesting density. GOSAT and GOSAT-2 provide more sparse and limited measurement

(16 days). This temporal dispersion may be insufficient to study small-scale emissions.

Although OCO-3 does not have a long history (as of 2019), due to the large number of observations, especially in urban areas, and as a result, good coverage of urban emission areas, high spatial resolution, and high accuracy, it is currently the most suitable satellite instrument in the study of CO_2 emissions in urban and local scales.

8.8 Future Trends

The measurement and study of XCO_2 by the remote sensing method with global coverage and long-term time history lead to a better understanding of the carbon cycle. However, the effects of geophysical factors such as airborne particles and clouds cause the dispersion of remote sensing data in space and time. For this reason, remote sensing measurements are sensitive to anthropogenic emissions on a regional scale, and this issue has been investigated less in researches [53].

Therefore, one of the useful solutions to solve the spatial and temporal variation of XCO2 data is to create XCO_2 maps (Mapping-XCO_2) using different satellite observations [53]. These maps interpolate the XCO_2 data from several satellites such as GOSAT and OCO-2 using geostatistical methods and solve the existing scatter and gaps. XCO_2 maps can be used to investigate the spatial and temporal changes of XCO_2 and study the carbon cycle [56]. In addition, the use of XCO_2 maps as a data set to calculate CO_2 emissions in urban scales has also been studied and found useful [53].

CO_2 emissions from urban areas are increasing worldwide. In order to implement measures to reduce CO_2 emissions at the regional, urban, and local scale, there is a need for more accurate and quantitative information on CO_2 emissions. To more accurately estimate CO_2 increase in local and point emission areas such as urban areas, power plants, and volcanoes, the method of removing the effect of concentration values in the nearby background area is proposed [16]. So through the difference between the XCO2 values at the emission source and the values in a clean area (as background) in the vicinity of the source, it is possible to estimate the amount of CO_2 emission at the production source [22]. An area where CO_2 emissions are negligible and not affected by pollutant emissions in its vicinity is called a clean area. Due to atmospheric disturbances, in order to estimate the amount of CO_2 emissions at the production source, the satellite observations of this area and the background area must have been taken almost at the same time.

The accuracy of CO_2 emission estimation by the remote sensing method is very sensitive to environmental conditions such as wind and atmospheric transport of CO_2. These data are affected by clouds, suspended particles, and weather conditions; and for this reason, they may not be of very good quality. For this reason, it is recommended to study the environmental conditions such as topography and wind condition of the area along with CO_2 emission estimation studies. Furthermore, validation with ground-based measurements such as XCO_2 measurements at TCCON stations can be useful to achieve more reliable results at local and urban scales [16]. Unfortunately, only a few cities around the world have ground CO_2 measuring stations. Therefore, satellite data with high accuracy and high spatial and temporal resolution, such as OCO-2 data as well as OCO-3 and GOSAT-2 data, are suggested for detailed and deeper studies [16].

Currently, CO_2 measurement in the SAM mode and the target mode by OCO-3 is an innovative dataset for urban and point scale carbon studies [22]. In order to more closely study the carbon cycle and CO_2 emissions, extensive space missions are planned in the future. With dense spatial coverage and short time repetition cycles, these spacecrafts can play an important role in quantifying CO_2 emissions in urban areas, potentially monitoring the effectiveness and progress of CO_2 emission reduction policies [22]. Table 8.5 presents a number of these space missions showing their applications and characteristics.

8.9 Conclusion

The increase in the population of cities has led to an enhancement of man-made CO_2 emissions and an increase in global temperature, especially in recent decades. By sending satellites that monitor greenhouse gases, monitoring the emission and measuring the CO_2 concentration have become more possible. In this research, the application of remote sensing technology to monitor CO_2 emissions in cities has been investigated. Therefore, CO_2 monitoring satellites (AQUA, ENVISAT, AURA, METOP-A, GOSAT, OCO-2, TANSAT, GOSAT-2, and OCO-3) were evaluated and compared in terms of anthropogenic CO_2 monitored in cities. This comparison is based on the parameters of species observed, CO_2 sensitivity, operational period, repeat cycle, spatial resolution, spectral range, CO_2 accuracy, observation mode, observation coverage, and observation density and gap. GOSAT, GOSAT-2, OCO-2, and OCO-3 satellites are effective in detecting CO_2 emissions at regional and urban scales due to the measurement of the entire XCO_2 column including areas close to Earth's surface. For studies that require a longer history, GOSAT can provide valuable information. The increase in CO_2 concentration in cities is between 0 and 6 ppm and generally less than 3 ppm. Therefore, observations of GOSAT-2, OCO-2, and OCO-3, which have a sampling accuracy of less than 1 ppm, can be used to monitor the amount of anthropogenic emissions in cities. OCO-2 and OCO-3 have much better spatial resolution than other satellites. Therefore, they provide more accurate measurements in detecting CO_2 emissions at urban and local scales. OCO-2 and OCO-3 cover more cities and more parts of these urban areas due to the large number of XCO_2 observations. The SAM observation mode on OCO-3 is specifically designed to measure anthropogenic emissions. This capability allows OCO-3 to survey the entire surface of a large city in a short period of time. Although OCO-3 does not have a long history of data acquisition (since 2019), due to the large number of observations, especially in urban areas, spatial resolution, and high accuracy, it is currently the most suitable satellite instrument in the study of CO_2 emissions at urban and local scales.

The use of remote sensing satellites for emission monitoring is only able to estimate the average column of the total mole fraction of CO_2 including the atmospheric region close to Earth's surface and cannot determine the amount of CO_2 gas emissions. However, they can monitor emission changes with high accuracy by removing the background. In recent years, many advances have been made in the field of satellite monitoring of greenhouse gas emissions, and many satellites have been placed in orbit for this purpose. In addition, in future missions, satellites

Table 8.5 Future planned space missions, their characteristics and applications [57]

Mission	Tansat-2	GeoCarb	Carbon mapper (CM)	MicroCarb	GOSAT-GW (TANSO-3)	Sentinel-Co_2M-A	CO_2Image
Instruments	CAPI2, LASHIS TANSAT-2 pollution	Scanning spectrometer (GeoCarb)	Hyperspectral imaging spectrometer	Microcarb	AMSR3-TANSO-3	CLIM, CO2I, MAO	COSIS
Agency	IAMCAS, SARI, STU	NASA	NASA, Arizona State University, JPL, ASU	CNES, UKSA	JAXA, MOE, NIES	COM-EUMETSAT	DLR
Launch date	2022	2022	2023	23-Dec	24-mar	25-Dec	26-May
Type	Sun-synchronous	–	–	Sun-synchronous	Sun-synchronous	Sun-synchronous	Sun-synchronous
Altitude (km)	600	35,400	–	650	666	735	500
Period (mins)	–	–	–		98.18	99.5	92.2
Repeat cycle (days)	–	1	–	7	3	11	–
Waveband	–	0.76, 1.61, 2.06, and 2.32 μm	–	10.5–17.6 nm	6.925–183.3 gh	405–2095 nm	~1.3 – ~3.0 μm
Spatial resolution	–	3×6 km^2	–	2×2 km^2	5–50 km	4 km^2	–
Accuracy	–	>2.7 ppm	–	1 ppm		0.7	–
Applications	6 satellite to measure CO_2, O2, CH4, and CO emissions	4-channel slit imaging spectrometer to measure XCO2, XCO, CH4, O2, and SIF	Pinpoint, quantify and track point-source methane and CO2 emissions.	Measure CO_2 concentration	Observation of GHG and study of water cycle mechanisms	Measure anthropogenic CO_2 emissions at country and megacity scales.	Measure CO_2 Emissions from localized sources

such as Tansat-2, GeoCarb, Carbon mapper (CM), MicroCarb, GOSAT-GW (TANSO-3), Sentinel-Co2M-A, and CO_2 Image can also play an important role in monitoring urban emissions in line with the policy of reducing CO_2 emissions.

References

1. UN-HABITAT, *Cities for All: Bridging the Urban Divide – State of the World's Cities 2010/2011 by UN-HABITAT* (2011)
2. R.K. Pachauri, L.A. Meyer, Climate change 2014: Synthesis report, in *Contribution of Work. Groups I, II III to Fifth Assess. Report of the Intergovernmental Panel on Climate Change* (2014)
3. L. Kamal-Chaoui, A. Robert, Competitive cities and climate change (2009)
4. K. Rypdal, W. Winiwarter, Uncertainties in greenhouse gas emission inventories – Evaluation, comparability and implications. Environ. Sci. Pol. **4**(2–3), 107–116 (2001). https://doi.org/10.1016/S1462-9011(00)00113-1
5. N. Bader, R. Bleischwitz, Measuring urban greenhouse gas emissions: The challenge of comparability. Cities Clim. Chang. **2**(3), 1–15 (2009)
6. L.R. Hutyra et al., Urbanization and the carbon cycle: Current capabilities and research outlook from the natural sciences perspective. Earth's Futur. **2**(10), 473–495 (2014)
7. G. Pan, Y. Xu, J. Ma, The potential of CO2 satellite monitoring for climate governance: A review. J. Environ. Manag. **277**(April 2020), 111423 (2021). https://doi.org/10.1016/j.jenvman.2020.111423
8. F.M. Bréon et al., *An Attempt at Estimating Paris area CO2 Emissions from* (2015), pp. 1707–1724. https://doi.org/10.5194/acp-15-1707-2015
9. Z. Zhongming, L. Linong, Y. Xiaona, Z. Wangqiang, L. Wei, *2019 Refinement to the 2006 IPCC Guidelines for National Greenhouse Gas Inventories* (2019)
10. J. Hakkarainen, I. Ialongo, J. Tamminen, Direct space-based observations of anthropogenic CO2 emission areas from OCO-2. Geophys. Res. Lett. **43**(21), 11400–11406 (2016). https://doi.org/10.1002/2016GL070885
11. R. Janardanan et al., *Comparing GOSAT Observations of Localized CO2 Enhancements by Large Emitters with Inventory-Based estimates*, no. June 2009, pp. 3486–3493, 2016. https://doi.org/10.1002/2016GL067843.Received
12. J. Hakkarainen, I. Ialongo, S. Maksyutov, *Analysis of Four Years of Global XCO 2 Anomalies as Seen by Orbiting Carbon Observatory-2* (2019), pp. 1–20. https://doi.org/10.3390/rs11070850
13. O. Schneising, J. Heymann, M. Buchwitz, M. Reuter, H. Bovensmann, J.P. Burrows, Anthropogenic carbon dioxide source areas observed from space: Assessment of regional enhancements and trends. Atmos. Chem. Phys. **13**(5), 2445–2454 (2013). https://doi.org/10.5194/acp-13-2445-2013
14. K. McKain, S.C. Wofsy, T. Nehrkorn, J. Eluszkiewicz, J.R. Ehleringer, B.B. Stephens, Assessment of ground-based atmospheric observations for verification of greenhouse gas emissions from an urban region. Proc. Natl. Acad. Sci. U. S. A. **109**(22), 8423–8428 (2012). https://doi.org/10.1073/pnas.1116645109
15. F.M. Schwandner et al., Spaceborne detection of localized carbon dioxide sources. Science **358**(80), 6360 (2017). https://doi.org/10.1126/science.aam5782
16. C. Park, S. Jeong, H. Park, J. Yun, J. Liu, Evaluation of the potential use of satellite-derived XCO2 in detecting CO2 enhancement in megacities with limited ground observations: A case study in Seoul using orbiting carbon Observatory-2. Asia-Pacific J. Atmos. Sci. **57**(2), 289–299 (2021). https://doi.org/10.1007/s13143-020-00202-5

17. S. Yang, L. Lei, Z. Zeng, Z. He, H. Zhong, An assessment of anthropogenic CO2 emissions by satellite-based observations in China. Sensors **19**(5), 1118 (2019)
18. A. Eldering et al., The orbiting carbon Observatory-2 early science investigations of regional carbon dioxide fluxes. *Science (80)* **358**(6360), 5745 (2017)
19. E.A. Kort, C. Frankenberg, C.E. Miller, T. Oda, Space-based observations of megacity carbon dioxide. Geophys. Res. Lett. **39**(17), 1–5 (2012). https://doi.org/10.1029/2012GL052738
20. G. Keppel-Aleks, P.O. Wennberg, C.W. O'Dell, D. Wunch, Towards constraints on fossil fuel emissions from total column carbon dioxide. Atmos. Chem. Phys. **13**(8), 4349–4357 (2013)
21. L. Lei et al., Assessment of atmospheric CO2 concentration enhancement from anthropogenic emissions based on satellite observations. Kexue Tongbao/Chin. Sci. Bull. **62**(25), 2941–2950 (2017). https://doi.org/10.1360/N972016-01316
22. M. Kiel et al., Urban-focused satellite CO2 observations from the orbiting carbon Observatory-3: A first look at the Los Angeles megacity. Remote Sens. Environ. **258**(August), 2021 (2020). https://doi.org/10.1016/j.rse.2021.112314
23. R. Beer, T.A. Glavich, D.M. Rider, Tropospheric emission spectrometer for the earth observing System's Aura satellite. Appl. Opt. **40**(15), 2356 (2001). https://doi.org/10.1364/ao.40.002356
24. A. Eldering, T.E. Taylor, C.W. O'Dell, R. Pavlick, The OCO-3 mission: Measurement objectives and expected performance based on 1 year of simulated data. Atmos. Meas. Tech. **12**(4), 2341–2370 (2019). https://doi.org/10.5194/amt-12-2341-2019
25. Y. Liu et al., The TanSat mission: Preliminary global observations. Sci. Bull. **63**(18), 1200–1207 (2018). https://doi.org/10.1016/j.scib.2018.08.004
26. T. Yokota et al., Global concentrations of CO2 and CH4 retrieved from GOSAT: First preliminary results. Sci. Online Lett. Atmos. **5**(1), 160–163 (2009). https://doi.org/10.2151/sola.2009-041
27. D. Wunch et al., Comparisons of the orbiting carbon observatory-2 (OCO-2) X CO2 measurements with TCCON. Atmos. Meas. Tech. **10**(6), 2209–2238 (2017)
28. R. Glumb, G. Davis, C. Lietzke, The TANSO-FTS-2 instrument for the GOSAT-2 greenhouse gas monitoring mission. Int. Geosci. Remote Sens. Symp., 1238–1240 (2014). https://doi.org/10.1109/IGARSS.2014.6946656
29. T. Hamazaki, Y. Kaneko, A. Kuze, K. Kondo, Fourier transform spectrometer for greenhouse gases observing satellite (GOSAT). Enabling Sens. Platf. Technol. Spaceborne Remote Sens. **5659**, 73 (2005). https://doi.org/10.1117/12.581198
30. D. Crisp, Measuring atmospheric carbon dioxide from space with the orbiting carbon Observatory-2 (OCO-2). Earth Obs. Syst. **9607**, 960702 (2015)
31. *GOSAT/IBUKI Data Users Handbook 1st Edition*, no. March (Satellite Applications and Promotion Centre, Space Applications Mission Directorate, Japan Aerospace Exploration Agency, Japan, 2011)
32. A. Kuze, T. Urabe, H. Suto, Y. Kaneko, T. Hamazaki, The instrumentation and the BBM test results of thermal and near-infrared sensor for carbon observation (TANSO) on GOSAT, in *Infrared Spaceborne Remote Sensing XIV*, vol. 6297, (2006), pp. 138–145
33. T. Hamazaki, Greenhouse Gases Observation from Space: Overview of TANSO and GOSAT (2008). https://doi.org/10.1117/12.2308255
34. EORC JAXA, TANSO-FTS – Fourier Transform Spectrometer. https://www.eorc.jaxa.jp/GOSAT/instrument_1.html. Accessed 18 Nov 2022
35. K. Shiomi et al., GOSAT level 1 processing and in-orbit calibration plan. Sensors Syst. Next-Generation Satell. XII **7106**, 71060O (2008). https://doi.org/10.1117/12.800278
36. *GOSAT-2/IBUKI-2 Data Users Handbook 1st Edition* (Satellite Applications and Promotion Centre, Space Applications Mission Directorate, Japan Aerospace Exploration Agency, Japan, 2020)
37. M. Nakajima, H. Suto, K. Yotsumoto, T. Miyakawa, and K. Shiomi, GOSAT-2: Development Status of the mission instruments, in *EGU General Assembly Conference Abstracts* (2015), p. 7731

38. Y. Oishi, H. Ishida, T.Y. Nakajima, R. Nakamura, T. Matsunaga, The impact of different support vectors on GOSAT-2 CAI-2 L2 cloud discrimination. Remote Sens. **9**(12), 1–13 (2017). https://doi.org/10.3390/rs9121236
39. Y. Oishi, T.Y. Nakajima, T. Matsunaga, Difference between forward-and backward-looking bands of GOSAT-2 CAI-2 cloud discrimination used with Terra MISR data. Int. J. Remote Sens. **37**(5), 1115–1126 (2016)
40. C.W. O'Dell et al., Improved retrievals of carbon dioxide from orbiting carbon Observatory-2 with the version 8 ACOS algorithm. Atmos. Meas. Tech. **11**(12), 6539–6576 (2018). https://doi.org/10.5194/amt-11-6539-2018
41. NASA/JPL, Orbiting Carbon Observatory-2. *NASA Jet Propulsion Laboratory*. https://ocov2.jpl.nasa.gov/. Accessed 20 Nov 2022
42. NASA/JPL, Orbiting Carbon Observatory-3. *NASA Jet Propulsion Laboratory*. https://ocov3.jpl.nasa.gov/. Accessed 20 Oct 2022
43. A. Eldering, T. E. Taylor, C. W. O. Dell, and R. Pavlick, The OCO-3 Mission; Measurement Objectives and Expected Performance Based on One Year of Simulated Data, vol. 3, no. June 2010 (2018)
44. D. Wu, J.C. Lin, T. Oda, E.A. Kort, Space-based quantification of per capita CO2 emissions from cities. Environ. Res. Lett. **15**(3) (2020). https://doi.org/10.1088/1748-9326/ab68eb
45. X. Ye et al., Constraining fossil fuel CO2 emissions from urban area using OCO-2 observations of Total column CO2. J. Geophys. Res. Atmos. **125**(8), 1–29 (2020). https://doi.org/10.1029/2019JD030528
46. D. Wu et al., A Lagrangian approach towards extracting signals of urban CO2 emissions from satellite observations of atmospheric column CO2 (XCO2): X-stochastic time-inverted Lagrangian transport model ('X-STILT v1'). Geosci. Model Dev. **11**(12), 4843–4871 (2018). https://doi.org/10.5194/gmd-11-4843-2018
47. J. Stoller-Conrad, Integrating carbon from the ground up: TCCON turns ten. Earth Obs. **26**(4), 13–17 (2014)
48. D. Wunch et al., Calibration of the total carbon column observing network using aircraft profile data. Atmos. Meas. Tech. **3**(5), 1351–1362 (2010)
49. G. Toon, J.F. Blavier, R. Washenfelder, D. Wunch, G. Keppel-Aleks, P. Wennberg, B. Connor, V. Sherlock, D. Griffith, N. Deutscher, J. Notholt, Total column carbon observing network (TCCON), in *Advances in Imaging. Optical Society of Ame* (2009), p. 2009
50. Y. Yuan, R. Sussmann, M. Rettinger, L. Ries, H. Petermeier, A. Menzel, Comparison of continuous in-situ CO2 measurements with CO-located column-averaged XCO2 TCCON/satellite observations and carbontracker model over the Zugspitze region. Remote Sens. **11**(24), 2981 (2019)
51. S. Oshchepkov et al., Simultaneous retrieval of atmospheric CO2 and light path modification from space-based spectroscopic observations of greenhouse gases: Methodology and application to GOSAT measurements over TCCON sites. Appl. Opt. **52**(6), 1339–1350 (2013). https://doi.org/10.1364/AO.52.001339
52. TCCON Data Archive, The TCCON Data Archive. https://tccondata.org/. Accessed 14 Oct 2022
53. M. Sheng, L. Lei, Z.C. Zeng, W. Rao, S. Zhang, Detecting the responses of co2 column abundances to anthropogenic emissions from satellite observations of gosat and Oco-2. Remote Sens. **13**(17) (2021). https://doi.org/10.3390/rs13173524
54. A. Kuze, H. Suto, M. Nakajima, T. Hamazaki, Thermal and near infrared sensor for carbon observation Fourier-transform spectrometer on the greenhouse gases observing satellite for greenhouse gases monitoring. Appl. Opt. **48**(35), 6716–6733 (2009). https://doi.org/10.1364/AO.48.006716
55. D. Crisp et al., The on-orbit performance of the orbiting carbon Observatory-2 (OCO-2) instrument and its radiometrically calibrated products. Atmos. Meas. Tech. **10**(1), 59–81 (2017)

56. M. Sheng, L. Lei, Z.C. Zeng, W. Rao, H. Song, C. Wu, Global land 1° mapping dataset of XCO2 from satellite observations of GOSAT and OCO-2 from 2009 to 2020. Big Earth Data, 1–21 (2022). https://doi.org/10.1080/20964471.2022.2033149
57. The CEOS Database, The European space agency, CEOS Database. http://database.eohandbook.com/database/missiontable.aspx. Accessed 16 Nov 2022

Correction to: The Potential of CO_2 Satellite Monitoring for Climate Governance

Fereshte Gholizadeh, Behrooz Ghobadipour, Faramarz Doulati Ardejani, Mahshad Rezaei, Aida Mirheydari, Soroush Maghsoudy, Reza Mahmoudi Kouhi, and Mohammad Milad Jebrailvand Moghaddam

Correction to:
Chapter 2 in: A. Ahmadian et al. (eds.), ***Carbon Capture, Utilization, and Storage Technologies,*** **Green Energy and Technology,**
https://doi.org/10.1007/978-3-031-46590-1_2

The (Rezaee) of [Mahshad Rezaei] was unfortunately published with an error. The initially published version has now been corrected.

The updated version of this chapter can be found at
https://doi.org/10.1007/978-3-031-46590-1_2

A. Ahmadian et al. (eds.), *Carbon Capture, Utilization, and Storage Technologies*, Green Energy and Technology, https://doi.org/10.1007/978-3-031-46590-1_9

Index

A. Ahmadian et al. (eds.), *Carbon Capture, Utilization, and Storage Technologies*, Green Energy and Technology, https://doi.org/10.1007/978-3-031-46590-1